THE NEUROSCIENCES INSTITUTE PUBLICATIONS SERIES

The Neurosciences Institute was founded in 1981 by the Neurosciences Research Program to promote the study of scientific problems within the broad range of disciplines related to the neurosciences. It provides visiting scientists with facilities for planning and review of experimental and theoretical research with emphasis on understanding the biological basis of higher brain functions.

The Institute has initiated an active publishing program. This volume, the first in a series of monographs by visiting scientists, outlines promising experimental strategies for studying neuronal circuits involved in higher brain functions.

Support for the Neurosciences Research Foundation, Inc., which makes the Institute programs possible, has come in part from generous gifts by The Vincent Astor Foundation, Lily Auchincloss, Francois de Menil, Sibyl & William T. Golden Foundation, Lita Annenberg Hazen, Lita Annenberg Hazen Charitable Trust, The IFF Foundation, Inc., Johnson & Johnson, Harvey L. Karp, John D. & Catherine T. MacArthur Foundation, Rockefeller Brothers Fund, van Ameringen Foundation, The G. Unger Vetlesen Foundation, and The Vollmer Foundation.

Publications

Neurophysiological Approaches to Higher Brain Functions
Edward V. Evarts, Yoshikazu Shinoda, and Steven P. Wise

Protein Phosphorylation in the Nervous System
Eric J. Nestler and Paul Greengard

Dynamic Aspects of Neocortical Function
Gerald M. Edelman, W. Einar Gall, and W. Maxwell Cowan, editors

Library of
Davidson College

NEUROPHYSIOLOGICAL APPROACHES TO HIGHER BRAIN FUNCTIONS

NEUROPHYSIOLOGICAL APPROACHES TO HIGHER BRAIN FUNCTIONS

EDWARD V. EVARTS
Laboratory of Neurophysiology
National Institute of Mental Health

YOSHIKAZU SHINODA
Department of Physiology
Tokyo Medical and Dental University

STEVEN P. WISE
Laboratory of Neurophysiology
National Institute of Mental Health

A Neurosciences Institute Publication
JOHN WILEY & SONS
New York   •   Chichester   •   Brisbane   •   Toronto   •   Singapore

599.01
E92n

Copyright © 1984 by Neurosciences Research Foundation, Inc.
Published by John Wiley & Sons, Inc.

All rights reserved, provided that parts of this work which were written as part of the official duties of an employee of the United States Government may not be copyrighted under the 1976 United States Copyright Act. Requests for further information should be addressed to the Neurosciences Research Foundation, Inc., 1230 York Avenue, New York, New York 10021.

Published simultaneously in Canada.

Library of Congress Cataloging in Publication Data
Evarts, Edward V.
   Neurophysiological approaches to higher brain functions.

   "A Neurosciences Institute Publication."
   Bibliography: p.
   Includes index. 85-2906
   1. Higher nervous activity. 2. Neurophysiology.
I. Shinoda, Yoshikazu, 1943– . II. Wise, Steven P.
III. Title. [DNLM: 1. Brain—Physiology. 2. Behavior—
Physiology. 3. Neural pathways—Physiology. WL 300
E92n]
QP395.E94 1984 599.01'88 83-25922
ISBN 0-471-80557-2

Printed in the United States of America
10 9 8 7 6 5 4 3 2 1

Preface

The Neurosciences Institute is dedicated to promoting an understanding of the nervous system in general and of higher brain functions in particular, and the present volume grew out of a Neurosciences Institute project directed toward developing strategies for studying neural circuits involved in a higher brain function. We will argue that "preparatory set," the capacity of an animal to elaborate plans for impending action, exemplifies a higher brain function amenable to neurophysiological study.

In order to formulate experimental approaches to the problem of preparatory set, we worked together at The Neurosciences Institute in 1982. It was apparent that it would be useful to consult with other investigators about the conceptual and methodological foundations of the experimental approaches that were being developed, and it was to this end that The Neurosciences Institute sponsored two conferences in the spring of 1982. The first conference was entitled "Dynamic Cerebral Circuits" and the second, "Behavioral Correlates of Identified Cell Types." In addition to the authors, both conferences were attended by Hiroshi Asanuma, Gerald Edelman, Leif Finkel, Einar Gall, Michael Merzenich, and Peter Zarzecki. The first conference was also attended by Stephen Lisberger and Nakaakira Tsukahara and the second conference included Eberhard Fetz, Edward Jones, and Vernon Mountcastle. These participants, some of the leading experts in fields ranging from neuroanatomy to neurophysiology and computer modeling of behavior, devoted their time and efforts to a critical discussion of many of the views presented in this monograph.

The volume that has emerged from this project consists of a survey of past work relevant to cells and circuits underlying preparatory set and the discussions held during the two conferences referred to above. There is no sense in which the present volume is meant to be a thorough review of any part of the scientific literature. Rather, it is an examination of what the neurophysiology of higher brain functions in mammals could be and an argument that it is practical and useful to study these questions with an emphasis on motor behavior.

The authors would like to thank Gerald Edelman, Director of The Neurosciences Institute, without whose encouragement this monograph and the conferences that formed its basis would not have been possible. We would also like to express our gratitude to the conference participants, listed above, with additional thanks to Eberhard Fetz, Edward Jones, and Peter Zarzecki, who commented on earlier versions of some of the chapters in this monograph. Finally, we are unendingly grateful to Betty Al-Aish, Deborah Bender, and Susan Wine for preparing the typescript and to Lori Budzinski for preparing the illustrations.

E. V. E.
Y. S.
S. P. W.

Contents

Chapter 1	**Neurophysiology, Behavior, and the Pyramidal Tract Neuron**	1

Stalking the Higher Function, 1
Plasticity versus Flexibility: Learning
   and the Use of It, 3
Pyramidal Tract Neurons, 5

Chapter 2	**Preparatory Set and Behavior**	7

The Term "Set" and the Importance of It, 7
From Leibnitz to Sherrington, 9
The Fall from Grace, 11
Mediating Neural Events in the Black Box, 12
Psychophysics to the Rescue, 15
Interactions Between Perceptual Set
   and Motor Set, 17
Switching Between Sets, 19

Chapter 3	**Set-Dependent Changes of Movement**	21

Set and Motor Output, 21
Comparison of Kinesthetic and Auditory
   Reaction Times, 27
The Question of Preprogramming, 29
Reflexes and Volitional Behavior, 30
Effects of Set on Automatic Responses, 33
Set Effects on Afferent Regulation of Speech, 34

Chapter 4	**Cerebral Cortex: Generalities and Specializations**	39

Behavioral Flexibility and the Cerebral Cortex, 39
Organization of the Cortex, 41
Cerebral Control of Movement, 44

Functional Specialization of the Precentral
Motor Cortex, 46

Chapter 5 **Cell Types and Information-Processing Circuits in the Cerebral Cortex**   49

Columnar Organization, 49
Inputs to Cortex, 51
Cortical Output Organization, 56
Neocortical Cell Types, 58
Intrinsic Cortical Organization, 59

Chapter 6 **Set and the Single Unit**   65

Wires in the Black Box, 65
Primitive Set-Related Neurons, 67
The Frontal Granular Cortex, 69
The Frontal Agranular Cortex, 78
The Parietal Somatosensory Cortex, 85
The Parietal Visual Cortex, 86
Temporal Visual Cortex, 87
The Auditory Temporal Cortex, 87
While a Monkey Waits, 89

Chapter 7 **Pathways for Set-Dependent Behavior**   91

Thalamic Inputs to Frontal Agranular Cortex, 91
Inputs to Thalamus from Globus Pallidus
   and Cerebellum, 96
Corticocortical Inputs: Parietal Cortex, 99
Electrophysiological Analyses of Interconnections, 103
Cerebellar Pathways to the Frontal
   Agranular Cortex, 103
Interpreting Field Potential Variation and
   Convergence of Cerebellocortical Pathways, 114
Cerebello-Thalamocortical Electrophysiology:
   Synthesis, 121
Pallido-Thalamocortical and Nigro-Thalamocortical
   Electrophysiology, 122
"Nonspecific" Thalamocortical Projections, 125

CONTENTS

Chapter 8 Experiments to Identify Pathways Involved in Set-Related Switching   127

    Correlations Between EPSPs and Field Potentials in Red Nucleus, 127
    Implications of Work on Structurally Based, Long-Term Adaptive Plasticity for Work on Dynamically Based, Short-Term Set, 131
    Experiments to Identify Brain Circuits for "Set," 134
    Gating in MI, 136
    Set-Dependent Gating of Cerebellar Signals in Thalamic Nuclei, 137
    Gating in Corticocortical Pathways, 139

Chapter 9 Behavioral Correlates of Electrically Evoked Field Potentials in the Cerebral Cortex   141

    Interpreting Field Potential Components, 141
    Modern Interpretations of Field Potentials, 142
    Electrically Evoked Field Potentials, 143
    Experimental Modification of Field Potential Components, 145
    Field Potentials and Short-Term Changes of Behavior, 147
    Use of Electrically Evoked Field Potentials in an Experiment on Set, 148

Chapter 10 Behavioral Correlates of Identified Cell Types in Cerebral Cortex   153

    Fiber-Tract Recording, 154
    Antidromic Activation Methods, 155
    Applications of Antidromic Identification Methods, 157
    Electrode-Sampling Bias Compensation and Antidromic Methods, 160
    Spike-Triggered Averaging, 164
    Laminar Analysis, 166
    Applicability of Neurophysiological Methods to the Identification of Cortical Cell Types, 169

Chapter 11 Summary and Speculation   171

    Summary, 171
    Speculation, 173

References   175

Index   193

Chapter 1

Neurophysiology, Behavior, and the Pyramidal Tract Neuron

STALKING THE HIGHER FUNCTION

Our thesis is based upon two observations which separately are fairly well accepted. The first is that the activity of output neurons in the primary motor area of the cerebral cortex is closely linked with the motor behavior of advanced mammals. The second is that the sensory signals which trigger that behavior are not: an identical sensory stimulus can trigger a variety of behaviors. All that we have to say revolves around these two propositions and an attempt to understand how the brain functions to control motor behavior and its associated neuronal activity in such a flexible manner. We believe such flexibility to be at the foundation of what are termed "higher brain functions" and its study to be of central importance to neurobiology.

Experimental psychologists have long proposed the existence of preparatory sets to explain how an identical stimulus can trigger one behavior at one moment and another behavior thereafter (Chapter 2). It has been inferred from many observations that there exists a set-dependent neural process for switching the behavioral consequences of a sensory stimulus depending upon the circumstances in which the stimulus occurs, and the effects of this set-dependent switching can be revealed by a careful examination of muscle activity (Chapter 3). We assume that the switching mechanism is mediated, at least partly, by the cerebral cortex (Chapter 4), and further, it seems to us almost inescapable that certain cortical output neurons (Chapter 5) in the primary motor cortex must be controlled, either directly or indirectly, by different inputs under different circumstances. This control may involve set-related neuronal activity (Chapter 6) and the pathways through which different motor control structures interact (Chapter 7). We seek to explore how a given sensory input might elicit a variety of behavioral patterns and how this might depend on preparatory set (Chapter 8). One approach involves recording simultaneously from a large number of neuronal elements (Chapter 9), and another is to examine the effect of set on input-output relationships of individual cortical projection neurons (Chapter 10).

In this monograph we discuss the neurophysiology of higher brain functions, but with an emphasis on motor processes rather than, for example, perception or learning and memory. Neurophysiologists interested in behavior have adopted a number of strategies. One approach is to find the basic building blocks of a behavioral repertoire and determine how these mechanisms work and how they might be assembled into a successfully functioning organism. As a recent example, it has been proposed (Gallistel, 1979) that behavior can be understood as the combination of three mechanisms: reflexes, oscillators, and servomechanisms. While we do not doubt that each of these mechanisms is important in the neural control of behavior, it seems unlikely that the higher-order functions of the nervous system can be inferred from such reduced models.

There are, of course, alternative viewpoints. Some neurophysiologists assert the importance of a detailed, if not complete, understanding of the neuromuscular synapse and the mechanics of muscle fibers before setting out on a more general study of behavior control. Others argue that before any understanding of supraspinal control of behavior is possible, the spinal and muscular systems must be thoroughly understood at a level far in advance of that currently available. And still others maintain that study of the cortex is unlikely to yield significant results until subcortical mechanisms are known with mathematical precision. We propose to attack the question in a different way. We would like to step back from the actual mechanisms involved in executing motor commands and from the various servomechanisms or obligatory reflexive mechanisms which operate the motor apparatus. We intend largely to ignore the mechanisms of movement execution and hope to gain a better understanding of higher brain function. Our approach is to "work backwards," to start from the behavior, specifically its motor aspects, and to look for the influences that act to select one of the several potential behavioral patterns that could occur under any given circumstance. In order to pursue such an objective, we feel that it is unnecessary to have a complete understanding of precisely how the motor apparatus interacts with neural commands to produce a movement. Rather, an empirical recognition, not understanding, of these distinct signals may be all that is required. It is our contention, then, that the study of brain circuits which lie at the basis of higher brain function is possible without a complete analysis of the "lower" functions.

Of course, there is great value in understanding the vegetative functions of the brain and the mechanisms of muscular control. But when one seeks to ask how an understanding of *higher* functions of the brain might be achieved, one can see that disappointingly little progress has been made to date. There are many currently applied approaches to this problem. As Geschwind (1980, pp. 663–664) states:

> . . . one approach is to continue to collect and describe [neurological] syndromes and study the lesions responsible for them. The results are of great usefulness to the clinical neurologist. One might study these syndromes from the psychologic or linguistic point of view, an approach that has been successful in the hands of such investigators as Benton, Hecaen, Milner, De Renzi, Vignolo, Zangwill, and many others. . . .
>
> However, a detailed study of neurologic mechanisms will, at least for the foreseeable future be possible only in nonhuman primates. It may be necessary for the more complex aspects of human behavior to be translated into simpler but still relevant behavioral paradigms in which nonhuman primates can be trained and thus be available for direct study. Consider, for example, the complex function of the frontal lobe in humans. It has been argued that this region contains neural mechanisms essential for foresight and socialization. It is obviously difficult or impossible to translate that concept into a combined behavioral-electrophysiologic experiment in the monkey. An alternative hypothesis is that the frontal lobe is responsible for the acquisition of complex responses to limbic stimuli (e.g., to anger, fear, sexual excitement, or hunger). This type of learning may underlie the socialization of the child, and it is at least possible that subhuman primates learn in a similar way the rules of their group for reacting to limbic stimuli. Although the translation to the primate training laboratory strains analogy, it is possible that it is only through such attempts that the neural mechanism of the truly higher functions can be brought under direct experimental observation.

We attempt here to set forth concrete approaches with available techniques that may lead to progress in understanding how the brain controls the most flexible behaviors of the organism. We think that an analysis of preparatory set will be a step toward understanding the neural control of these behaviors and that set may be a window to higher brain functions.

PLASTICITY VERSUS FLEXIBILITY
LEARNING AND THE USE OF IT

We focus on the neurophysiology of higher brain functions because of the profound adaptive advantages that these functions confer on mammals. Learning is an ancient ability of nervous systems, and we emphasize that our focus is not on the mechanism of learning, but rather is on some of the neural circuits that can utilize information already learned to execute a behavior. Here one can contrast *neural plasticity*, and other processes that may underlie information storage, with a process that might be termed *neural switching*.

The importance to the organism of neural switching can be appreciated readily. As Arbib (1981, p. 1459) points out: "The intelligent organism does not so much respond to stimuli as to select information that helps it achieve current goals." If, as this view suggests, an animal prepares for stimuli and actively searches its environment for them, it seems likely that, when encountered, these stimuli trigger a response preplanned for that stimulus. In this way, the animal could make use of the vast amount of stored information and interact flexibly with the environment. Mechanisms of this kind may be elaborated to allow a given stimulus to guide behavior under a variety of circumstances, depending on the expectations, or "set," of the animal; hence the necessity for some sort of neural switching mechanism. In fact, the very detection of a sensory stimulus depends on set, a process termed selective attention: the ability of animals to direct their cognitive apparatus toward specific aspects of their environment (internal or external). This may mean that olfactory cues as opposed to visual stimuli will be used to guide behavior, or it may mean stimuli in one part of the visual space or one aspect of visual information rather than another will be used. In any case, the principle is the same: the animal "expects" a certain class of stimuli and prepares to receive them. The term "perceptual set" is employed for this concept and is discussed in more detail in Chapter 2.

The situation is closely analogous in the motor system. To be optimally efficient and rapid in responses to stimuli, the animal should be set to make a certain pattern of movements. It has been shown many times that, under such circumstances, reaction time (defined as the time from the receipt of a movement-triggering stimulus to the time the neural command reaches the effector apparatus) is substantially shorter when a subject knows what movement must occur (see Chapter 3).

One way in which this process might occur involves central representations of stimuli and motor acts. A centrally generated representation of a type of stimulus or a specific stimulus that might be encountered (see, e.g., MacKay and Gardiner, 1972) would not only facilitate recognition of the stimulus and its interpretation, but would also be expected to produce recognition errors when a sufficiently similar stimulus occurs. In parallel, we can imagine a central representation of a motor response pattern, a pattern that facilitates the execution of a response to the appropriate stimulus. This use of the term "representation" is not to be confused with a topographic representation. The latter is a temporally stable feature of connectivity and selective processes within a sensorimotor system, and when changes occur they are gradual; the former is a dynamic process that is short-lived. We refer to the dynamic neural representation of the motor pattern as the "motor set." This concept is also more fully addressed in Chapter 3, but note here that by arranging the behavioral control system so that an expected stimulus can be linked to the execution of a planned motor response, an extremely efficient and rapid reaction can be generated to almost any stimulus. Such a system would require only that any expected stimulus could be linked with any planned behavior to give it the flexible characteristics observed in mammalian behavior under natural conditions. Thus, set can be considered the process by which a movement is planned and a stimulus hypothesized.

PYRAMIDAL TRACT NEURONS

In studying a higher brain function as described above, one must take into account the organizational complexity of the vertebrate nervous system and choose the site for investigation carefully. The primary motor area of the cerebral cortex and its output system, the pyramidal and corticospinal tracts, has many advantages as an initial locus of study. The pyramidal tract system, the fibers of which connect cell somata in the cortex with the brainstem and spinal cord, is closely linked with the motor apparatus and originates from the neocortex, the presumed site of higher functions. Further, it offers a number of technical advantages over other cortical fiber systems.

Two technical advantages stand out: the relatively large size of many of the cell somata which send axons down the pyramidal tract, and the compactness of the tract. The latter feature means that the cell bodies of these axons in the lower brainstem can be readily identified by the technique of antidromic axonal activation (see Chapter 10). Electrical stimulation of pyramidal tract elicits action potentials in the large cell bodies of pyramidal tract neurons, which can be monitored with extracellular microelectrodes. There are many advantages related to cell size, an important practical consideration for any single-unit recording method. Larger neurons produce relatively large electric fields when they discharge, and these fields can then be monitored at a greater distance from their source than can smaller neurons. Importantly, substantial progress has been made in understanding the interaction of the microelectrode and pyramidal tract neurons of various sizes (see Chapter 10), making feasible an estimation of the activity of the entire population of these cells in a given region. Two additional features of the pyramidal tract confer important advantages for neurophysiological analysis. All of the pyramidal tract fibers originate from cell bodies situated within one, and only one, of the six layers of the neocortex. The advantage conferred by the nearly unilaminar arrangement of pyramidal tract neurons is important for single-cell recording, and the subsequent interpretation of their location in the cortex. And finally, the central feature of primary motor cortex that makes it attractive for our purposes: the activity of these neurons is already known to be very closely correlated with motor behavior (Evarts, 1981, and Chapter 4).

Chapter 2

Preparatory Set and Behavior

One approach to identifying circuits for higher brain function involves the systematic manipulation of preparatory set as an experimental variable. Such manipulation takes advantage of the fact that the response to a given stimulus depends upon prior instructions that determine the preparatory set. The importance of an instruction stimulus (IS) as a variable in a behavioral experiment was pointed out by Postman (1963), who suggested that the IS is a necessary part of the antecedent conditions in any perceptual experiment. In his view, the IS is an independent variable that can control the readiness of the subject to respond selectively. The IS can be thought of as "setting" the subject so that a subsequent trigger stimulus (TS) then elicits one of a number of different responses ($R_1, R_2, \ldots, R_n$) depending on the prior IS ($IS_1, IS_2, \ldots, IS_n$). In this chapter we review results of some past experiments and ideas on set.

THE TERM "SET" AND THE IMPORTANCE OF IT

Gibson (1941) has pointed out that the term "set" has been used as an explanatory concept for so many related but separable phenomena that its meaning has become blurred, and has suggested that the term should always be qualified. A similar point of view was adopted by Ryan (1970), who listed a number of different ways in which the term set has been defined:

1. Differences of emphasis in performance of the same task, as, for example, when the subject must respond with either speed or accuracy.
2. Readiness for a stimulus.
3. Aftereffects of a *previous* stimulus or previously executed task.
4. Intention of the subject to perform a particular task.

Our use of the term set relies most heavily on the last of these definitions, but also to a large extent on the second.

Woodworth (1958) believed that the concept of set is useful and even essential in describing behavior, but also concluded that the word set as it is often employed has too many meanings. As a more restricted and formal definition Woodworth (1958, p. 41) proposed that:

> ... preparatory set is a state of readiness to receive a stimulus that has not yet arrived or a state of readiness to make a movement that cannot be done until a preliminary movement has been made.

For simplicity, we would alter Woodworth's definition by deleting the last phrase. Thus, we define set as *a state of readiness to receive a stimulus that has not yet arrived or a state of readiness to make a movement*.

One way to make the term set less inclusive is to use "selective attention" in reference to sensory input processing, and to reserve the term set for use in reference to movement. Hebb (1972) adopted this distinction between set and attention. Having defined a *mediating process* as a dynamic activity of neurons that can hold the excitation delivered by a sensory event after the event has ceased, Hebb (1972, p. 84) went on to note that:

> ... the mediating process that does the holding is apt to introduce selectivity into the behavior, in either or both of two ways, in the form of *attention* and *set*. Accordingly, these also are marks of higher behavior. Attention is selectivity in what is responded to, or sensory selectivity; set is a selectivity of response, motor rather than sensory. Very often, however, attention and set go together.

Though Hebb used different terms to refer to these two sorts of mediating processes, he clearly appreciated that motor set and selective attention (i.e., perceptual set) are closely linked conceptually and are essential to a consideration of higher brain functions. Accordingly, this chapter deals with both concepts, especially since: the ideas that have emerged from studies of selective attention in perceptual tasks are highly relevant to an understanding of motor set; and observations relevent to motor set (e.g., reaction time, movement speed, and movement accuracy) have been used in studies on selective attention.

Our view of the importance of preparatory set is similar to that of Schoenfeld and Cumming (1963, p. 233) who noted that:

> ... although instructions to the subject have been an almost universal part of psychophysical and perceptual experiments, they are sometimes overlooked in systematic treatments.
>
> Instructions pose a special problem for several reasons. It is not clear that they have the status of stimuli at any point in the experiment other than at

the time when they are actually given by the experimenter. Even at that time, there may be no clearly observable response to them. After the experiment is begun, they are frequently not repeated again, and it may be questioned whether a stimulus given before an experiment is, in any ordinary sense, acting as a stimulus in the later stages of the experiment. Moreover, only very infrequently is the subject's specific conditioning history with respect to the instructions known at all. In some studies, the experimenter may give a few "practice trials" to make sure that the subject "understands the instructions," but often even this step is omitted. . . .

Thus, when told to "cancel all words that contain both the letters *e* and *s*," the human adult may need to present these instructions to himself repeatedly while performing the task, until the complex discrimination is sufficiently well trained to short-circuit self-instruction in this form. A task for human subjects may be greatly complicated by changing the instructions on each trial. Even simple alternation of instructions may initially produce chaotic behavioral consequences.

Thus, understanding of set is essential for a full description of animal behavior. While this view may seem obvious now, it has not always been so.

FROM LEIBNITZ TO SHERRINGTON

The history of the concepts involving set is rich, both in the number of treatments of these topics and the antiquity of the concepts themselves. Boring (1957) and Herrnstein and Boring (1965) provide full descriptions of the development of the concepts from antiquity to the emergence of experimental psychology as a scientific discipline in the second half of the 19th century, including accounts of the ideas of Plato, Aristotle, Descartes, Hobbes, Locke, Berkeley, Hume, Mill, and Spencer. Accordingly, we do not attempt to review their ideas here. Berlyne (1969) deals specifically with the development of the concept of attention in psychology and notes that perhaps the starting point for modern ideas on set and attention was the observation of Leibnitz that many external stimuli excite sense organs without reaching conscious experience. But the research of Donders and Kulpe (see Ladd and Woodworth, 1911) marks the dawn of the current era of experimental work on set. Donders showed the relationship of task complexity to the speed of reaction and led to much of the subsequent work on motor set; Kulpe performed studies that showed the effects of expectancy on perception, thereby laying the groundwork for subsequent work on perceptual set.

Ladd and Woodworth (1911, p. 470) give an excellent account of the status of thinking on set and attention at the beginning of this century:

Many thousands of experiments have been made (since the work of Donders in 1868), with the use of the most complicated and delicate machinery, in order to fix the amount of time required for the various processes, both nervous and mental, which are the conditions of our conscious life. These experiments have succeeded in bringing many interesting facts to light. But the laws thus established beyond all reasonable question are remarkably few; moreover, they are nearly all merely restatements in more definite form of already familiar generalizations. That a kind of sluggishness or inertia, which the stimulus must overcome, belongs to all the senses, and that they often continue to act, when once roused, after the exciting cause is withdrawn; that different sensations following each other too quickly tend to confuse or destroy each other; that no one can see or think more than about so rapidly, but that this rate varies with different individuals and with the same individual at different times; that it takes more time to perceive or think where the objects are complex, and are either too small or too large or too closely alike; that it takes time to will or choose, less time to act when we know what to expect, and more time to move, in response to a particular sensation, some part of the body which we are not accustomed to connect with that sensation; that practice increases the speed of our mental and bodily action, and that fatigue and certain drugs diminish it—all these statements were matters of common observation long before experimental psychology began its use of scientific methods.

If "many thousands" of psychological experiments had already been done by 1911, then how many more must have been carried out in the subsequent seven decades?

But ideas on mediating processes such as set and their effects on stimulus-response relations in the late 19th and early 20th centuries were not provided entirely by experimental psychologists. Indeed, Boring (1957, p. 31) points out that contributions to experimental psychology during the early part of the 19th century were made primarily by men who ". . . did not think of themselves as psychologists nor of their subject-matter as psychology. They were physiologists, physicists or astronomers. There were no scientists who styled themselves psychologists until well after 1860."

Among early *physiological* studies of the effects of set on stimulus-response relations, those of Sherrington are especially relevant. Indeed, it was his work concerning the properties of reflexes that first demonstrated how a change in central excitatory state could profoundly modify the reflex responses to identical stimuli. Sherrington (1933, pp. 14–15) called attention to the profound state-dependent flexibility of the response to a sensory stimulus:

> . . . the same stimulus can from occasion to occasion call forth different results, even amounting to actual reversal. The reflex stimulus which at one time extends the limb at another will flex it. Such actual reversals are traced on analysis to a different ratio of weighting so to say of the nerve-nets with

inhibition and excitation. A previous stimulus leaves the ratio of inhibition to excitation for a time altered. The external stimulus itself proves always to be excitatory and inhibitory admixed. The outside world through certain receiving stations is always generating inhibition in a number of nerve-nets, just as it is generating excitation in others. A drug may in a second wholly reverse a reflex by tilting the inhibition-excitation balance in an adjoining nerve-net. A little more or a little less of inhibition or conversely of excitation on a nerve-net and the pattern of the reflex shifts like the pattern of a tapped kaleidoscope.

At about this time, however, the scientific outlook toward behavioral effects mediated by unseen "states" or "sets" markedly changed.

THE FALL FROM GRACE

Attitudes toward set underwent a dramatic change with the coming of what Berlyne (1969) refers to as the "behaviorist revolution," which was followed by a fall from grace of unobservable, hypothetical mediating processes such as perceptual set (i.e., attention). The principal reasons for this change in attitude are listed by Berlyne (1969, p. 2) as follows:

1. The study of attention had been treated as inseparable from the study of consciousness. Attention had been discussed either as a process that determines what we are conscious of or else (e.g. in Titchener's writings) as an attribute ("clearness") distinguishing some experiences from others. Consequently, those who wanted psychology to concentrate on publicly observable behaviour and to eschew introspection were induced to throw attention out with the bath water.
2. The notion of attention seemed to imply a homunculus. Attention was apparently conceived as something inside us that "decided" whether or not we would accept the influence of a particular stimulus. In fact, one variety of attention that had long been recognized was "voluntary" attention, which seemed to stir up all the recalcitrant problems surrounding volition and perhaps even free will. This seemed quite unacceptable to those who were seeking laws that related responses to stimuli.
3. Above all, the concept of "attention" seemed unnecessary. Objectively, we manifest attention to a particular external stimulus by performing a motor response that corresponds to it and can be attributed to its causal influence. Surely the familiar laws of behaviour fully suffice to determine how powerfully a stimulus determines the response. It may evoke an unlearned response because of some innate reflex or other inherited behaviour pattern. Alternatively, and more frequently in higher animals, it may evoke a response because that response has become associated with it through prior learning.

MEDIATING NEURAL EVENTS IN THE BLACK BOX

In spite of the negative attitudes toward set that developed in the 1930s, a number of psychologists continued to be interested in these concepts. One of these was Donald Hebb, who gradually succeeded in rehabilitating the concept and ushering in the modern era of work on set. Hebb proposed a number of informal neural models, models seeking to formulate events in the black box between stimulus and response, and we will soon consider these models. Since Hebb drew on the earlier formulations of William James, we first consider one of the ideas of this early psychologist. James's diagram of a neural model of set is shown in Figure 2.1, where interneuron K responds either to sensory cell S or to feedback from the muscles. James proposed that neural paths become more effective with repeated use, and that with such repeated use the capacity of sensory neuron S to directly excite interneuron K rises. Further, he proposed that when S elicits activity in K, its activity corresponds to the *idea* of the movement. As James (1890, p. 585) wrote: ". . . translated into psychic terms: *when a sensation has once produced a movement in us, the next time we have the sensation, it tends to suggest the idea of the movement, even before the movement occurs.*" James's circuit may be thought of in the present context as one underlying a *long-term* perceptual set for the receipt of input to K upon the occurrence of activity in S. However, set in the sense that it is now commonly used refers to an evanescent, rather than to a long-term, process, so we will now consider

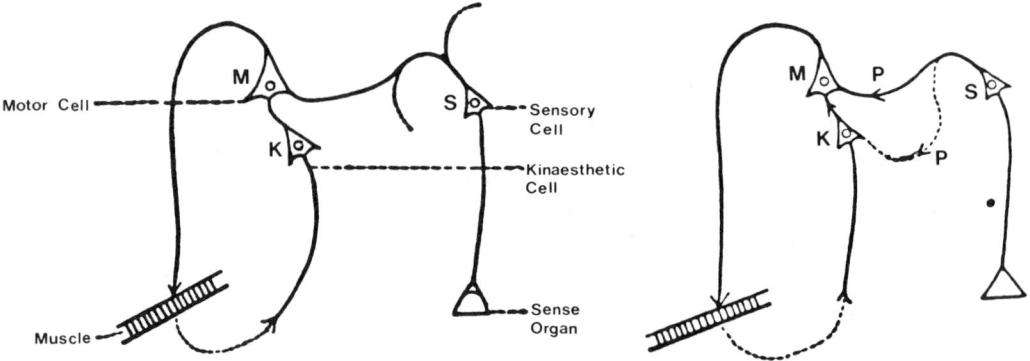

Figure 2.1. *A neuronal circuit for the idea of movement*. Left: activation of a sensory cell S reflexly excites a motor cell M, resulting in muscular contraction and feedback via the kinaesthetic cell K. Right: an additional pathway (the dashed line P) is envisaged; this pathway represents the circuit for an associative memory. James proposed that when the sensory cell S is excited from the periphery it will discharge into K as well as into M. As a result, K will be excited by S even before it receives kinaesthetic feedback. (From James, 1890)

some ideas on neural events underlying abrupt changes of set and involving short-term changes in levels of brain activity. We hasten to point out that changes in set need not be thought of as involving exclusively one or another of these mechanisms. There is abundant evidence that changes of set, at least in certain uses of the concept, involve a mix of long-term and short-term processes. But here we concentrate on short-term phenomena and consider proposals that short-term changes of tonic brain activity may play a role in set-related changes of behavior. One of the most influential theories involving tonic neuronal discharge was proposed by Hebb on the basis of physiological and anatomical evidence for circuits in which thalamic nuclei projecting to a given zone of cerebral cortex receive return signals from these same cortical areas. In describing his informal model, Hebb (1959, p. 628) stated that:

> The key conception is that of the *cell assembly*, a brain process which corresponds to a particular sensory event, or a common aspect of a number of sensory events. This assembly is a closed system in which activity can "reverberate" and thus continue after the sensory event which started it has ceased. Also, one assembly will form connections with others, and it may therefore be made active by one of them in the total absence of the adequate stimulus. In short, the assembly activity is the simplest case of an *image* or an *idea*: a representative process. The formation of connections between assemblies is the mechanism of association.

Thus, by Hebb's time, speculation concerning the inside of the black box had come "out of the closet." Hebb also pointed out that assemblies could be formed in association with occurrence of motor activities, but that in many cases, the activities of these assemblies would not be observable externally. He proposed that these representational cell assemblies, each corresponding to some property of environmental stimulation, would form associative links with each other and with motor-related cell assemblies. Associations would be established between specific sensory modalities and related motor activities (e.g., visual stimulation with eye movements and somatosensory stimulation with movement of the corresponding part of the body). To quote Hebb (1959, pp. 628–629):

> . . . each [sensory modality] would therefore establish neural connections with, and tend to produce, its own motor activity. However, actual muscle contraction would often not occur, because some other assembly activity occurring at the same time might inhibit the motor path, or simultaneously active assemblies might have motor effects that were physically incompatible with each other (e.g., flexing and extending a limb at the same time). Overt movements would result whenever such inhibition or conflict was absent.

Hebb also sought to diagram the role of cell assemblies in the higher functions of the nervous system. He defined the mediating process, the basis of higher function (1972, p. 84):

> ... as an activity of the brain which can hold the excitation delivered by a sensory event after this event has ceased, and thus permit a stimulus to have its effect at some later time. To "mediate" means to form a connecting link, and the simplest function of the mediating process is to connect S with R. Theoretically, however, a mediating process can also be excited by other mediating processes as well as its own sensory event, and when a number of mediating processes interact in this way—being excited by each other as well as by sensory events—the result is thinking; so, theoretically, a mediating process might also be defined as the unit or elementary component of thought, replacing the term "idea."

These ideas culminate in Hebb's (1972) simple neural model of short-term set (shown in Figure 2.2) in which identical stimuli elicit very different responses, depending on prior instructions. It illustrates the case of a person

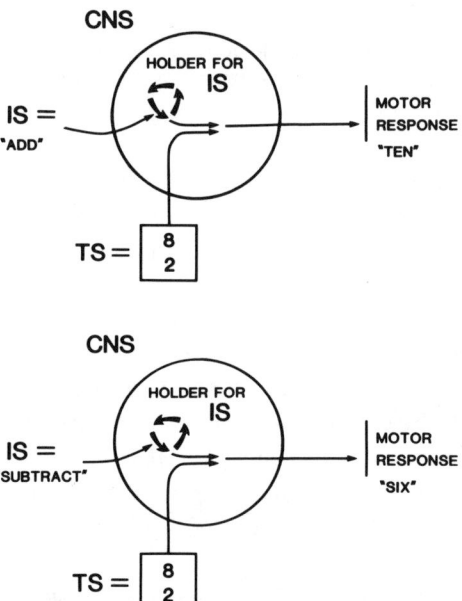

Figure 2.2. *A simple neural model of set.* An "addition set" following an IS (in this case an auditory stimulus) to "ADD" is contrasted to a "subtraction set" following an auditory IS to "SUBTRACT." Depending on set, the same subsequent TS (8, 2) delivered via visual stimulation would then evoke a different response. In this diagram, the excitation from the prior IS is held in short-term memory and the TS (8, 2) is connected with one of two motor responses ("ten" or "six"), depending on the subject's set to add or subtract.

who is first instructed to say the sum of two numbers and a few seconds later is given the two numbers. Alternatively, the person might have been told to say the difference between two numbers, and with this "subtraction set" the same two numbers would then evoke a different response.

This discussion of Hebb's view of preparatory set concludes the first part of our historical introduction. It should now be evident that the ideas of motor and perceptual set have a long history. Unfortunately, in our view, neurophysiologists have paid scant attention to them. It seems to us, however, that Hebb's formulations can serve as an important foundation to future physiological study of the higher functions hidden inside the black box of the central nervous system. With the advances that have occurred since these ideas were proposed, neural models have become more explicit and formal, and there are sometimes objections to terming Hebb's ideas models, since they lack any parametric specifications (e.g., the strengths, signs, and weightings to be assigned to interacting circuits). Thus, when speaking of the ideas of Hebb as a model, we are not using the term in any strict sense. It is difficult to categorize the various models or formulations that have evolved in the past century but, roughly speaking, one can place models in one of three classes. The first class, exemplified by the models of Hebb and of James, is informal and descriptive. A second class, exemplified by the models of Broadbent (1958) for selective attention and Sternberg et al. (1979) for motor set, are more specific but still lack the parameter specifications that have become the hallmarks of a third category exemplified by the models of Cooper (1981), Anderson (1977), and Edelman (1978, 1981; Edelman and Reeke, 1982). We believe that models in all three categories provide useful ideas for neurophysiological experiments.

PSYCHOPHYSICS TO THE RESCUE

In the 1940s, a number of factors combined to lead to the resurrection of research on "set," "attitude," and "expectancy" within the field of experimental psychology. Results provided by neurophysiology (e.g., electroencephalographic changes with visual attention in the absence of changes of overt behavior) provided a central nervous system correlate of attention. In addition, the emergence of information theory and engineering psychology and the growth of applied psychology during World War II led to precise specification of effects of vigilance, set, and selective attention in operators of military equipment. These developments, in turn, led to entirely new branches of psychology and neurophysiology, branches that have evolved in the aerospace research laboratories and applied psychology and physiology laboratories whose investigators have made numerous contributions to current understanding of attention and performance. As a result of these changes, the concept of set, in disfavor during the behaviorist revolution, was rehabilitated. With this return to respectability came a new generation of experiments based on information theory, psychophysics, human engineering, experimental and physiological psychology, and neurophysiology. The results of this new generation of experiments, together with a few of their antecedents, are considered in the remaining parts of this chapter.

As noted above, a powerful impetus to studies on the role of set in perception and movement was provided by progress in electrical engineering and information theory during the 1940s and 1950s. Swets (1964) notes that theories concerning signal detection as provided by mathematicians and engineers, though constructed without regard for human sensory processes, nevertheless provided a general theory that is of use in understanding human detection and recognition behavior.

Some of the most precise information on effects of set on signal detection have resulted from studies in which prior information as to the pitch of a sound to be presented has been shown to modify the detectability of the sound. Swets (1964) noted that by the mid-1950s it became apparent that there were several reasons for considering the role of central, or cognitive, factors in auditory frequency analysis. At this time, it seemed that rather than viewing the auditory detection process as being based upon the operation of fixed sensory elements, it would be more fruitful to view the system as one in which varying central factors modified the processing of information arising from sensory receptors.

Though ideas such as those referred to above are widely accepted now, the period prior to 1950 was one in which there was a tendency to view sensory reception as being based on fixed sensory receptor processes. The changes in the prevailing point of view can perhaps be attributed to the neuropsychology of Lashley (1951), Sperry (1952), and Hebb (1949, 1955). Later, an important reappraisal of the role of central factors in sensory processes was provided by the work of Broadbent (1958), in which it was shown that the operator's ability to process information clearly pointed out the importance of the concept of selective attention.

One of the major issues that has arisen in studies of the effects of attention and set on signal detection concerns the extent to which observed effects of prior instruction depend upon a perceptual mechanism or, alternatively, whether a response process might be responsible. Swets and Sewall (1964) investigated the effects on signal detection of different ways of providing information about the stimulus to the subject. Effects of information given prior to the response were inferred by changes in the proportion of correct responses, denoted P(c). It was already known that when uncertainty about the frequency of an auditory stimulus exists, P(c) is lower than when the signal frequency is known. Swets and Sewall studied the roles of perceptual versus motor set. In their experiments, the signal that the subject was required to detect was a tone of either 500 or 1100 Hz (0.1 sec duration) presented on a continuous background of white noise. They first compared P(c) when information about stimulus frequency was given before or after the presentation of the tone. The frequency information (cueing) was supplied by two lights: the left light indicated that the stimulus was the lower of the two frequencies and the right light meant that it was the higher frequency. In another experiment, cueing was with tones instead of lights, thus examining the effects of cueing by samples of the tones themselves. It was found that cueing after the auditory stimulus failed to enhance performance. Thus, providing the subject with information as to the frequency of the signal *after* the presentation of the signal and noise was not beneficial. In contrast, *prior* cueing had beneficial effects, although the effects of visual cueing were considerably weaker than those of cueing by tones.

The experiments comparing cueing before versus after a stimulus bear on the question of whether the beneficial effect of cueing depended on enhancement of a motor process or a perceptual process, and were based on the assumption that a cue given after a stimulus but before the response to the stimulus could enhance the motor process as well as or better than the same cue given before the stimulus. If one accepts this assumption, the finding that prestimulus cueing enhances tone detection but poststimulus, preresponse cueing fails to do so means that the effect of cueing depends on a perceptual, rather than on a motor, mechanism. In the context of our interest in set, it is also of note than even when cueing by tones immediately preceded the test stimuli, the overall performance of subjects was not as good as when an *entire block* of trials was given with all signals being of the same frequency.

INTERACTIONS BETWEEN PERCEPTUAL SET AND MOTOR SET

The data showing effects of set on perceptual, rather than on motor, processes were obtained by using what Moray (1970) has referred to as the puritan stimuli of tone bursts and psychoacoustics. Such stimuli are commonly used in experiments aimed at investigating selective attention, and the concept of perceptual set was introduced to refer to a concept that has much in common with the older term selective attention, but that lacks the introspectionist taint that attention had acquired during the behaviorist revolution. Paired with the concept of perceptual set was that of motor set. A common theme of psychophysical experiments has been a determination of the extent to which a given set-dependent behavior may depend on perceptual set as compared to motor set. For example, when a subject listens to two simultaneous messages (a different message for each of the two ears) and attempts to repeat one of the messages, the effects of set involve both the perceptual and the motor levels. The experimental paradigm of presenting one message to the left ear and a different message to the right ear with the prior instruction to pay attention to the message delivered to one ear and ignore the one presented to the other ear is referred to as dichotic listening. In the experiments of Cherry (1953) on dichotic listening, the subject was asked to repeat the message delivered to the specified ear as quickly as possible after its presentation. Cherry, who called this repetition *shadowing*, found that subjects could shadow a message delivered to one ear and ignore a message delivered to the other. Whereas Swets and Sewall (1964) had concluded that the set for detection of pure tones was at the level of perception, rather than motor response, the results of Cherry and others on shadowing showed that both perceptual and motor response sets were involved.

To see why this is so we should consider a most important difference between dichotic listening experiments and more traditional psychophysical signal detection experiments, in which the only required response was "yes" when the signal was detected, with no response at all if it was undetected. Broadbent (1970) noted that, in contrast, the Basic English used in dichotic listening tasks contained 850 words, and repetition of a two-word sentence might call for any one of 732,500 "brain states." A three-word sentence might require over 60 million states and a consequent increase in complexity of the response process. This added complexity led to greater interest in designing paradigms to distinguish perceptual set from motor set. Dichotic listening experiments (Cherry, 1953; Cherry and Taylor, 1954; Broadbent, 1958, 1970) showed, not surprisingly, that as the response complexity increased, the ability of the subject to detect the stimulus was compromised. Thus there appears to be an interaction between perceptual and motor set. Moray (1970) notes, along these lines, that if a person is trying to write a message, he will not be able to report on the content of a different message presented to the ears. One way to explain this observation, in the current context, is to postulate a limitation on the number of motor sets that can be established at one time.

SWITCHING BETWEEN SETS

Later (Chapters 7 and 8) we consider circuits that link cerebellum to motor cortex via thalamus and speculate on switching of motor cortex control from one cerebellar output (interpositus nucleus) to another (dentate nucleus) with a change in motor strategy (from maintenance of stability to rapid

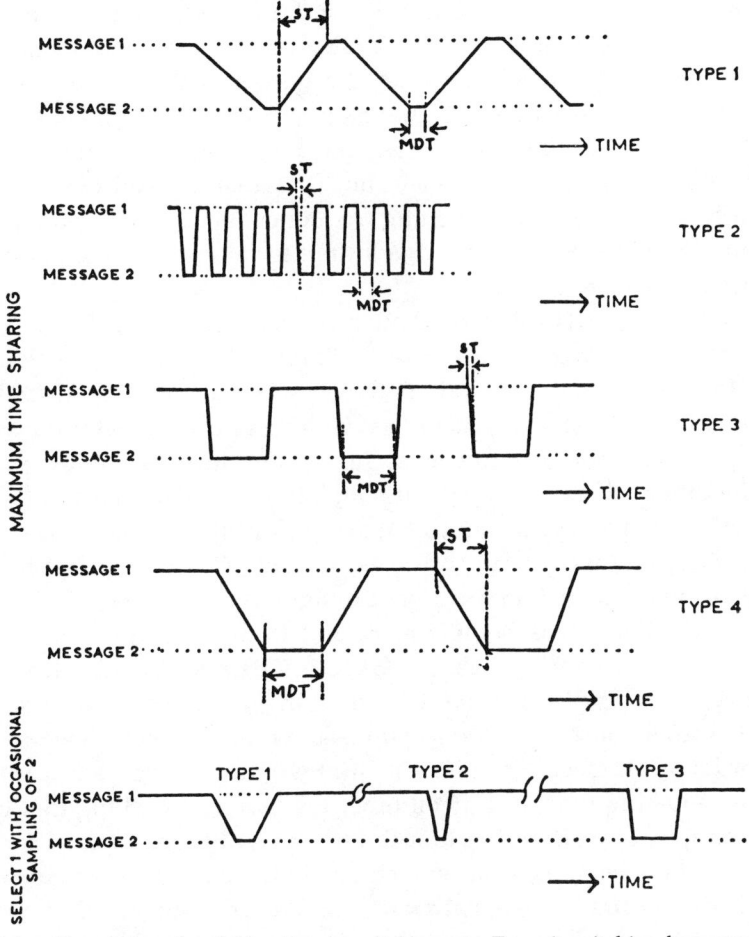

Figure 2.3. *The relation of switching time to switching rate.* Type 1 switching has a very short minimum dwell time (MDT) but a long switching time (ST), giving very poor time-sharing performance. Alternate relations between dwell time and switching time are shown in types 2, 3, and 4. Actual psychophysical data have not yet firmly established the values for switching time in dichotic listening experiments, but several estimates suggest a switching time of about 50 msec and a dwell time of about 100 msec. (From Moray, 1970)

movement). In these speculations we are concerned with how quickly a switch in motor set can be effected. Although we have little information on the processes involved in switching motor set, some data exist on the time taken to switch attention. In asking "How long does it take to switch attention?" two aspects of the switching process should be considered. First, one must be concerned with how long it takes to get from one attentional set to the other. This changeover time, when the subject is neither in one attentive state or the other, may be viewed as a "dead time," such as exists in a mechanical switch while the common contact is traveling from one fixed contact to the other. Second, one must be concerned with the minimum dwell time after attention has just entered a particular state. Together, dead time and dwell time determine switching frequency, i.e., the number of times per second that a subject can alternate between two attentional foci. Moray (1970) has noted that a long dead time must obviously produce a low switching frequency, but that a short dead time need not necessarily produce a high switching frequency, because it is possible that a system may have a minimum dwell time for attention to remain in the state that it has just entered.

Cherry and Taylor (1954) carried out an experiment in which they sought to determine how long it takes to shift channels by periodically switching speech from one ear to the other. Slow switching rates produced little difficulty. However, at fast switching rates it was impossible for the subject to shift attention from ear to ear, so he merely devoted attention to one ear or the other. By measuring the intelligibility of the speech as a function of switching frequency, a point of maximum difficulty was found when the switching rate was about three per second. This meant that the time interval during which "nothing of the signal is perceived at either ear because the 'attention is being transferred from one ear to the other'" (Cherry and Taylor, p. 557) is about 1/6 sec or 170 msec. Additional evidence (Moray, 1970) suggests that the dead time involved in a switch from one ear to the other can be as little as 50 msec, whereas dwell times must be on the order of 100 msec. Figure 2.3 illustrates various hypothetical relations between switching time and dwell time. Of course, switching that occurs when a subject pays attention first to one ear and then to the other must differ in certain respects from switching that occurs when motor cortex output is driven first by one pathway and then by another. Nevertheless, there is enough in common between models for dichotic listening and those for motor set switching to permit the comparison.

Chapter 3

Set-Dependent Changes of Movement

SET AND MOTOR OUTPUT

In the previous chapter we discussed some of the major points in the historical development of the concept of set, and implied that there may be an intimate relationship between perceptual set (i.e., selective attention) and motor set. Further, we showed how a change in perceptual set affects the ability to detect and process information coming in through various channels. The phenomena termed perceptual set are observable because they have clear-cut behavioral consequences. That is, the ability to detect signals or to process incoming information can be measured by the performance (often inferred from verbal report) of an experimental subject in a behavioral task. In the same way, motor set can be inferred to have existed by aspects of the movements that follow it, even though motor set in itself should produce no directly observable result.

Much of our understanding of how set affects motor output has been derived from measurements of reaction time (RT). It has long been known that RT is shorter when subjects have gotten "set" to respond as a result of having advance knowledge of the movement that is to be elicited by a forthcoming trigger stimulus (TS). The term "simple RT" is used in reference to reaction time that occurs when the subject has this advance knowledge. Often there is only one motor response throughout an entire block of trials. Such a block might involve sequences in which:

1. The subject depresses a telegraph key.
2. A "ready signal" is delivered after a delay of several seconds.
3. A TS is delivered after an additional delay (one to two seconds).
4. The subject releases the key as rapidly as possible after the TS.

In this situation, the subject needs no instruction as to the required motor response for each individual trial, as the required motor response is the same throughout the entire block and a single instruction at the beginning of the block is sufficient. In a variant of the simple RT procedure, the subject knows in advance what movement must be made in response to TS, but acquires this knowledge for each trial separately, rather than for an entire block of trials. The subject's new motor set for each successive trial is established by an instruction stimulus (IS) that specifies the movement to be made when the TS is delivered. An example of such a variant might involve sequences in which:

1. The subject positions a handle in a midposition such that it can be moved in either of two possible directions.
2. An IS is delivered after a delay of one to two seconds, specifying the direction of movement that is to be made upon delivery of the TS.
3. The TS is delivered after another delay of one to two seconds.
4. The subject moves the handle in the IS-specified direction as soon as possible after TS.

"Choice RT" differs from simple RT by virtue of the fact that the TS not only triggers movement but, at the same time, specifies which one of a number of possible alternative movements is to be executed; in this case, the TS is also an IS. The RT that occurs when a subject is required to move in a direction indicated by one of several lamps as soon as it is illuminated is called choice RT, which is longer than the simple RT.

Information theory was applied to RT studies in the 1950s, and it was proposed that choice RT increases in proportion to the logarithm of the number of alternative choices. However, Mowbray and Rhoades (1959) showed that sufficient practice could virtually eliminate this effect of the number of alternative choices. Figure 3.1 shows the experimental apparatus that was used for this study, in which one subject performed a grand total of 15,000 trials over several months. It was found that with sufficient practice there was no difference between two-choice and four-choice RT, and that differences between one-choice (i.e., the simple RT) and ten-choice RT became very small. Effects of practice have been noted by many investigators, and Schneider and Shiffrin (1977) proposed a two-process theory of human performance, one process called controlled and the other automatic. Schneider and Shiffrin (1977, p. 1) suggested that, with sufficient practice, the automatic process proceeds:

> . . . without subject control, without stressing the capacity limitations of the system, and without necessarily demanding attention. Controlled processing

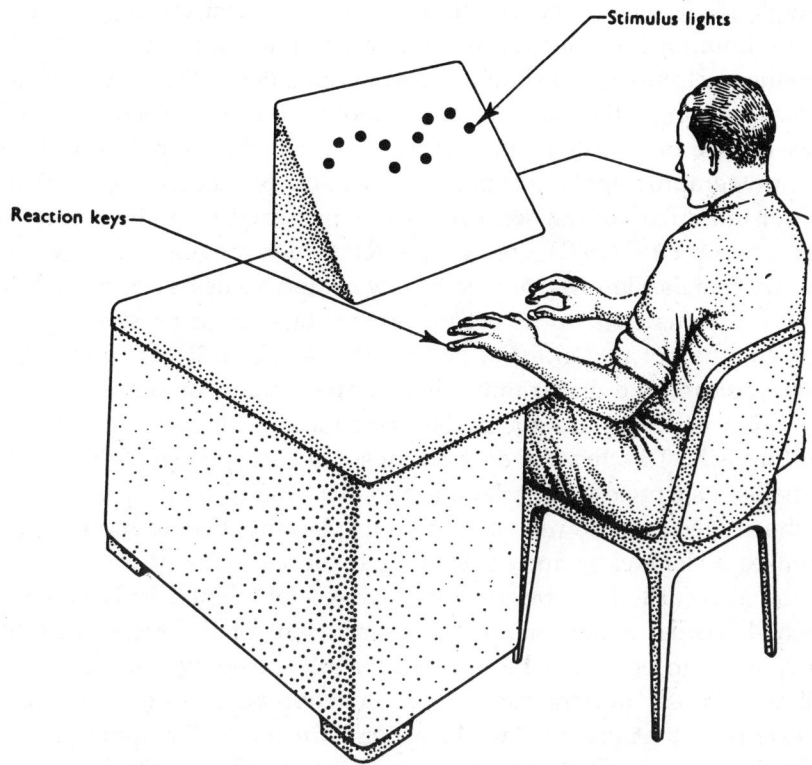

Figure 3.1. *Apparatus for a ten-choice reaction time experiment.* This apparatus was used to test the hypothesis that practice might modify the proportional relation between choice reaction time and the logarithm of the number of alternative stimuli. When subjects practiced extensively with an apparatus such as the one shown above, this proportional relationship disappeared. (From Mowbray and Rhoades, 1959)

is a temporary activation of a sequence of elements that can be set up quickly and easily but requires attention, is capacity-limited (usually serial in nature), and is controlled by the subject.

Subsequent studies have shown that RT becomes longer in proportion to the logarithm of the number of alternative responses under certain conditions, but that when subjects have had enough practice to become automatized there is much less lengthening. Thus, motor sets appear to be established with experience in a given situation and these provide substantial RT improvements, even in a choice RT paradigm. For a review of these findings and of the issues involved in interpreting them, the reader is referred to Shiffrin and Schneider (1977) and Schneider and Shriffrin (1977).

Simple RT in well-practiced subjects has long been taken as a measure of the minimum time needed to initiate voluntary muscular contraction. When the electromyogram (EMG) is used as an index of the start of muscular contraction, simple RTs are commonly about 150 msec. When such indices as key-release or handle displacement are used, the lag between the first electromyographic sign of muscular activity and detection of the mechanical result of muscular contraction causes the measured RT to increase. Thus, the review by Fitts (1951) cites simple RT values ranging from 230 to 500 msec; much of the slowness and variability in these values is due to differences in the apparatus used to detect movement onset. For present purposes, however, it is not necessary to go into the details of RT experiments; we merely wish to note that commonly accepted values for simple RTs (from TS to onset of EMG) are about 150 msec for auditory TSs and about 180 msec for visual TSs. In view of the widespread acceptance of these values, it came as a surprise when Hammond (1956) found that a prior IS which told the subject how to react to a limb displacement could modify muscle responses at latencies much briefer than the latencies of the simple RT. Hammond recorded the biceps EMG evoked by a limb displacement that stretched biceps, and found that a component of the response (with a latency of 50 msec) could be profoundly influenced by a prior IS. When the IS was "resist," and the subject was set to oppose the externally produced arm extension that stretched the biceps muscle, an EMG response occurred at 50 msec. Such a response was absent, however, after an IS to "let go," which had set the subject to allow the externally produced arm extension to proceed without any interference. The biceps tendon jerk (20 msec latency) did not change with these two different ISs in Hammond's experiment. The time from the stretching TS to the divergence of the two motor responses that depended on the IS was 50 msec when the response-indicator was EMG activity and 73 msec when the response-indicator was force, with the increased latency for force resulting from the lag between the first electrical signs of EMG activity and the detection of the force changes due to muscle contraction. That these differences due to the IS "resist" versus the IS "let go" occurred with a latency about half of voluntary RT suggested to Hammond (1956) that "a stretch reflex is at work when the arm is forcibly extended." Hammond added, however, that "this must be reconciled with the fact that prior instructions to 'let go' can interfere so rapidly and effectively with the subject's response." Hammond's apparent uncertainty as to how the set-dependent phase of biceps-muscle response was to be viewed resulted from the fact that the intended, IS-dependent muscle activity (beginning at 50 ms) was superimposed upon the stretch reflex (which begins at 20 ms), and so it was difficult to separate precisely the reflex from the nonreflex components of biceps activity.

Evarts and Granit (1976) and Evarts and Vaughn (1978) sought to make a clearer delineation between the reflex and nonreflex components of displacement-evoked biceps activity in an experiment designed so that a given direction of intended arm movement was triggered by either of two possible directions of limb displacement; one direction reflexly excited and the other direction reflexly inhibited the musculature that was to become active with the intended response that a prior IS had set the subject to make. When the triggering displacement shortened and "unloaded" the muscle the subject was set to activate, for example the biceps, the initial reflex of TS was biceps inhibition, and the subsequent biceps discharge arose from the silence of the unloading reflex. This sequence allowed the time of onset of the IS-dependent intended discharge to be accurately determined, because it took place *in spite of*, rather than because of, the reflex effects of the TS. Results obtained in this experiment showed that arm displacements could trigger intended muscle discharge at latencies as short as 70 msec, even when the initial reflex effects of the TS were inhibitory.

In these experiments, subjects grasped a handle that could be rotated by pronation or supination of the forearm and maintained the handle in a vertical orientation as they awaited the IS, which was followed at a variable interval by the TS. Figure 3.2 illustrates the experimental apparatus. The handle grasped by the subject was coupled to the axle of a torque motor that could generate steady-state forces requiring that the subject maintain tonic muscular activity (pronation or supination) to position the handle vertically. By regulating the current through the motor, the experimenter could control the steady-state activity of supinating (e.g., biceps)

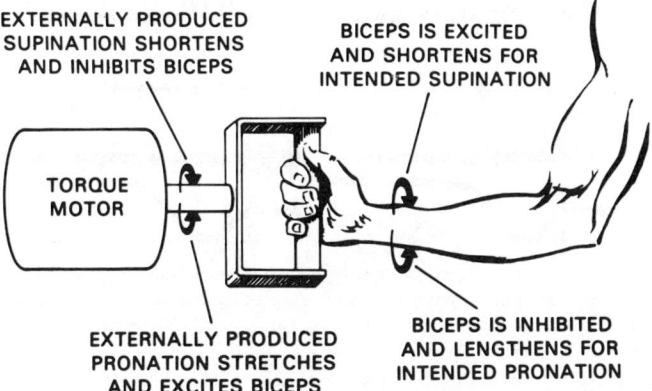

Figure 3.2. *Experimental apparatus for triggering intended movements with kinesthetic stimuli. Handle movements generated by the torque motor could either supinate (shortening and inhibiting the biceps muscle) or pronate (stretching and exciting biceps). For active supination, biceps was excited, whereas biceps exhibited inhibition when the subject actively pronated his forearm. (From Evarts and Granit, 1976)*

and pronating muscles. After the correct handle position was maintained for a period of 2–5 sec, one of two possible IS lamps was illuminated. One IS lamp set the subject to respond to a subsequent TS by pronating the handle; the other set the subject to respond by supinating the handle. Thus, the intended pronation or supination movement did not occur in response to the IS itself, but was elicited by the subsequent TS that displaced the subject's hand. The interval between the IS and the TS varied unpredictably between 1.8 and 2.5 sec.

Figure 3.3 illustrates intended biceps discharge triggered at a latency of 70 msec by a displacing TS that shortened the biceps passively, thereby causing an initial reflex silence of the tonic discharge that had been present before the arm displacement. The lower part of the figure shows trials in which the reflexly inhibitory TS was again delivered, but this time after an IS that had set the subject to make a movement involving biceps silence, so that both the reflex and the intended responses of the biceps involved quiescence. The 70-msec latency discharge of the biceps after the unloading reflex, as shown in this figure, is dependent on the set of the subject and is analogous to the intended EMG discharge with RT movements triggered by auditory or visual stimuli. Many subjects without prior experience in

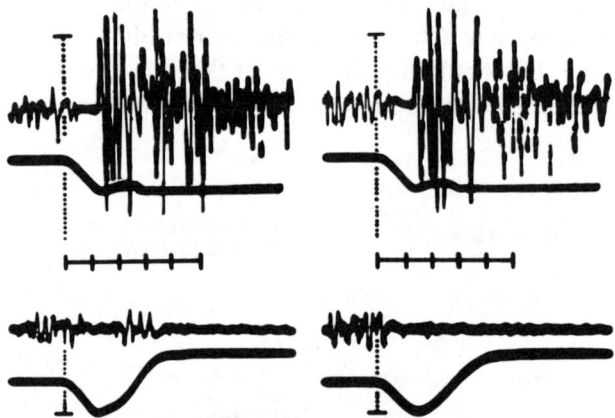

Figure 3.3. *Muscle discharge triggered by a reflexly inhibitory limb displacement.* In all four sets of traces, biceps was tonically active when an externally produced displacement (beginning at the dotted line) supinated the arm, shortening biceps and eliciting an inhibitory "unloading reflex" at a latency of 25 msec. Top: a prior IS had set the subject to supinate, and the biceps discharge necessary for the supination overcame the reflex inhibition at a latency of 70 msec. Bottom: a prior IS had set the subject to pronate, and biceps remained virtually silent following TS. The heavy trace below each EMG record indicates handle position, with supination being indicated by downward deflection and pronation by upward deflection. Time marks are at 50 msec intervals. (From Evarts and Granit, 1976)

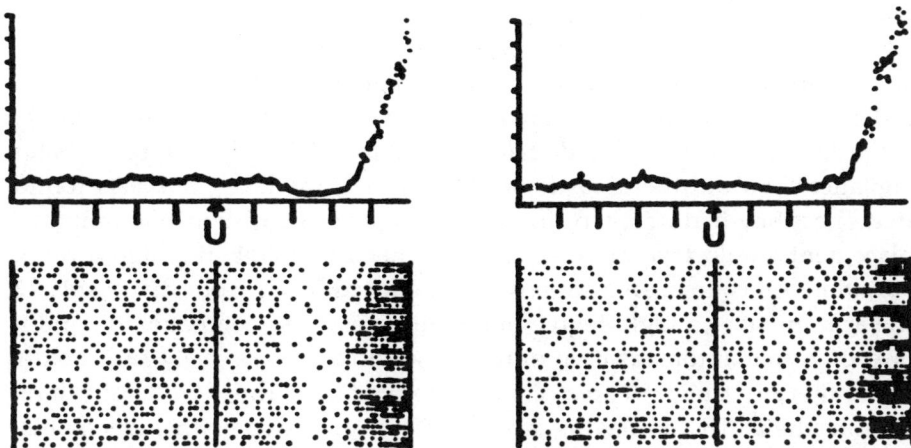

Figure 3.4. *EMG rasters and average responses to a kinesthetic TS involving muscle shortening.* The rasters and average responses shown above illustrate the variability of the RT (as measured by EMG activity) when TS was triggered by an inhibitory unloading (U) stimulus. Results for two different subjects are illustrated at the left and right. Time marks are at 25 msec intervals, and raster and average response displays represent muscle activity occurring 125 msec before and after the kinesthetic TS. It is apparent in the raster displays that for many trials no EMG response at all was present even at the extreme right-hand margin of the raster (125 msec following TS). For the responses at the right, 10 of 32 trials failed to show an increase in EMG activity even at 125 msec. The shortest latency responses in this particular subject occurred at approximately 85 msec. (From Evarts and Vaughn, 1978)

this kinesthetic RT task found it rather difficult at first to execute intended movements that "contradicted" the reflex responses to the triggering displacement, as demonstrated in the upper part of Figure 3.3. Coupled with this difficulty were longer and more variable latencies of the intended discharge, as shown in Figure 3.4 Thus, the 70-msec latency shown in Figure 3.3 is a value obtained with a well-practiced subject who had become "automatized" in the sense used by Schiffrin and Schneider (cited above). In spite of this automatization, however, such a well-practiced subject could abruptly shift from a set for turning a muscle on to a set for turning it off.

COMPARISON OF KINESTHETIC AND AUDITORY REACTION TIMES

It has been pointed out that kinesthetic RTs can be as short as 70 msec even with reflexly inhibitory TSs, and that RTs are considerably longer when the TS is an auditory signal. Kinesthetic and auditory RTs also differ with respect to variance. Figure 3.5 shows the results of an experiment in which RT movements were triggered by randomly interspersed kinesthetic and auditory stimuli. The auditory RTs, many of which were greater than 250 msec, were much more variable than the kinesthetic RTs. Furthermore, subjects commonly reported that the auditory RT task required greater vigilance than did the kinesthetic RT task. For the latter, subjects found that once they had adopted the set to supinate, the response to limb displacement would take place without the need to maintain a high level of attention. Of course, an arm movement elicited by an input that displaces the arm involves a high degree of stimulus-response compatibility, and it is known that automatization of responses occurs more easily and rapidly when there is a high degree of such compatibility. Thus, we might think of kinesthetic TSs as having a "special" relationship with movements, presumably reflecting the simple fact that the sensory and motor systems are in this case intimately related. The 150-msec value we have spoken of as typical for auditory RT was obtained in experimental set-ups that failed to maximize stimulus-response compatibility, because the input was to the ear and the output was from the arm. If, however, the amplitude of the kinesthetic stimulus is decreased sufficiently (Figure 3.6), the apparent benefit of stimulus-response compatibility disappears and the kinesthetic RTs begin to resemble auditory RTs in both latency and variance. Here we can consider the relationship between the low-amplitude somatosensory

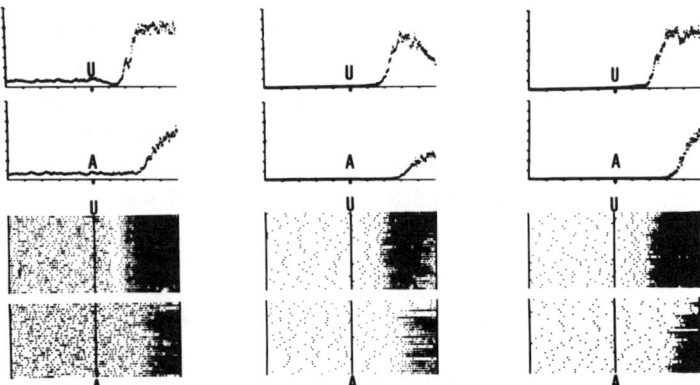

Figure 3.5. *Comparison of auditory and kinesthetic reaction times.* Results for three subjects are shown in the three columns above. Rasters and average responses depict muscle activity in an RT experiment with a kinesthetic, unloading (U), or auditory (A) TS. The displays depict EMG activity for 250 msec before and after the TS. EMG responses occur approximately 50 msec earlier for the unloading TS (U) than for the auditory TS (A), and RT variability is much greater for the auditory than for the unloading TS. (From Evarts and Vaughn, 1978)

THE QUESTION OF PREPROGRAMMING

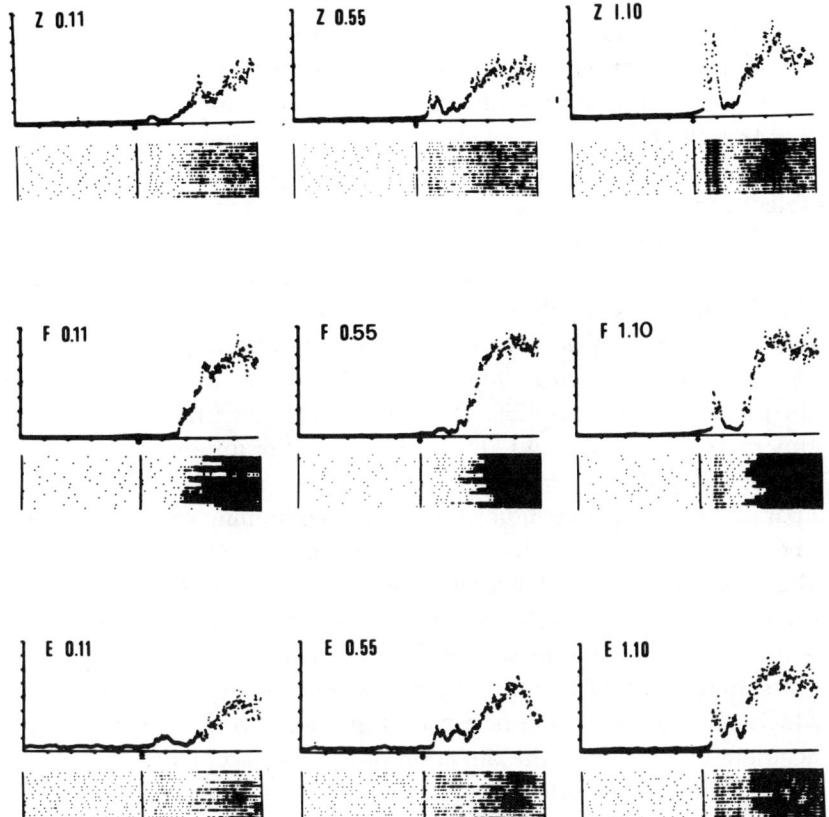

Figure 3.6. *Effects of kinesthetic TS intensity.* EMG rasters and average responses were obtained from three subjects at three different TS intensities. Each row of rasters and average responses corresponds to one of the subjects (Z, F, and E). The initial reflex response increases greatly as stimulus intensity is increased from 0.11 ft-lb (left) to 1.10 ft-lb (right), and two of the subjects (Z and E) also showed a decrease in response latency for the intended component with increasing stimulus intensity. (From Evarts and Vaughn, 1978)

stimulus and movement to be a "general" one, in which the shortest-latency EMG responses to a length change are not observed.

THE QUESTION OF PREPROGRAMMING

For the displacement-triggered arm movements that we have been considering, the subjects had 1.8 to 2.5 sec to get set to promote or supinate between the IS and the TS. In order to determine the extent to which short kinesthetic RTs depended upon the subject having the opportunity to get set for a particular movement, a different pair of ISs was used. The new ISs meant that the subject's movement should be *opposite to* or *in the same direction* as the triggering displacement. These ISs might be considered to be of a higher order than pronate or supinate, since same or opposite does not give the subject an instruction about his next movement in such a precise manner as pronate or supinate. Instead, the higher-order IS gives the subject partial information to which must be added the TS for a final determination of the proper movement to be executed. In this paradigm, the successive arm displacements were randomly interspersed, so that even after the IS the subject could not know whether to supinate or pronate in response to the TS. EMG latencies for intended biceps discharge were more variable and of longer average latency with these same or opposite ISs than with the pronate or supinate ISs. This finding agrees with the observations of both Crago et al. (1976) and Houk (1976) that minimum kinesthetic RTs are 70 msec when the impending movement could be preprogrammed, and that a paradigm change precluding such preprogramming caused kinesthetic RT to become 10–50 msec longer. But even these longer values for choice kinesthetic RT are still shorter than those for choice visual or auditory RT. It is thus apparent that a relatively short kinesthetic RT does not require the establishment of a motor set for the specific movement which is to be made, but that some RT benefit may be obtained from a more general set: one which calls for activation of muscles that will oppose or assist a limb displacement even though the direction of the displacement is not known in advance. Similarly, RTs for eye movements decrease by 60 msec or more when removal of a fixaation point signals that a saccade target will be presented at a later time (Fischer and Boch, 1983).

From the standpoint of our interest in effects of set on movement, the studies considered above have provided at least some delineation between reflex and set-dependent EMG responses to limb displacement. The reflex responses begin at a relatively constant latency of about 20 msec, whereas the set-dependent responses have less constant latencies, with minimum values of about 50–70 msec even for rapid stretches of the sort used by Hammond. Later, we consider the pathways that are involved in these two components of EMG activity and describe the underlying set-dependent and reflex motor cortex discharges.

REFLEXES AND VOLITIONAL BEHAVIOR

The notion that set is a higher brain function that provides a basis for behavioral flexibility leads to the question of reflexes and volitional behavior.

A motor behavior that is associated with a higher-order function is often termed volitional; lower-order functions are termed reflexive. A neural reflex, in its conceptually simplest form, is an obligatory coupling between a sensory stimulus and a motor response. A classic example is the myotatic reflex, in which a muscle stretch induces a certain amount of muscle contraction through a spinal mechanism. A volitional or voluntary movement, also in its conceptually simplest form, is a movement initiated within the brain (almost always assumed to be in the higher "centers" of the brain). Though heuristically useful, it is probably a mistake to rigidify these concepts or to consider the voluntary movements as an entity. Indeed, J. Hughlings Jackson (1931) recognized long ago that the processes underlying volitional movement are subject to the laws of reflex action and that reflexes and volitional movements are not opposites. However, if volitional movement cannot be defined by exclusion (i.e., as something that is not reflex), then how can it be defined? Granit (1977) concluded that what is volitional in voluntary movement is its purpose. A related opinion was expressed by Bernstein (1967, p. 133):

> If the program of a motor act is discussed, as a whole, in macroscopic terms we cannot discover any other determining factor than the image or representation of the result of the action (final and intermediate) in terms of which this action is directed and which leads towards the comprehension of the corresponding motor problem.

Thus, Jackson, Bernstein, and Granit suggested that the volitional features of a voluntary motor act are to be thought of in terms of the goal or result toward which this act is directed, but that the actual events underlying acquisition of the goal are built up from a variety of reflex processes. Gurfinkel (personal communication) has also defined volitional movement in relation to the goal and, in studies of the electromyographic and kinematic characteristics of expert marksmen, he found that a pistol remains virtually immobile in spite of the fact that many parts of the body exhibit movement. The key to pistol stability was that for each movement of the trunk or limbs, a corresponding counterbalancing movement stabilized the position of the pistol in space. Many reflex systems were involved in achieving this goal. Gurfinkel's example of the marksman recalls the views of James (1890, p. 497):

> The marksman ends by thinking only of the exact position of the goal, the singer only of the perfect sound, the balancer only of the point of the pole whose oscillations he must counteract. The associated mechanism has become

so perfect in all these persons that each variation in the thought of the end is functionally correlated with the one movement fitted to bring the latter about. Whilst they were tyros, they thought of their means as well as their end: the marksman of the position of his gun or bow, or the weight of his stone; the pianist of the visible position of the note on the keyboard; the singer of his throat or breathing; the balancer of his feet on the rope, or his hand or chin under the pole. But little by little they succeeded in dropping all this supernumerary consciousness, and they became secure in their movements exactly in proportion as they did so.

If a volitional movement is to be defined in terms of a "conscious" goal (to use these terms loosely) and whatever mechanisms are associated with achieving that goal, how, then, can we define a reflex? Perhaps reflexes can be defined as innate stimulus-response circuits, i.e., although they may be modifiable, they do not depend upon experience with a stimulus for their existence. Let us begin by considering the cough reflex, the corneal reflex, and the vestibulo-ocular reflex, all of which are adaptively advantageous to an animal: the cough reflex maintains an unobstructed airway, the corneal reflex protects the eye, and the vestibulo-ocular reflex keeps a relatively stable image on the retina. But, in our subjective experience, none of these effects are conscious goals. For our purposes, we might follow J. Hughlings Jackson (1931), who introduced the term "automatic" as a substitute for the reflexive. But rather than speaking of automatic and nonautomatic movements, Jackson spoke of movements that were more or less automatic. Jackson recognized that a given movement becomes more automatic with practice. And just as volitional movements can become more automatic with practice, so it is possible for well-practiced subjects to make their reflexes less automatic, sometimes to a remarkable extent. For example, the "gag reflex" that most individuals experience the first time a tube is passed through the nasopharynx into the esophagus can be overcome with practice. Students participating in experiments that require frequent passage of such a tube to sample stomach contents can learn to introduce the tube without gagging at all. Similarly, one can mimic the gag reflex voluntarily with practice. Further, individuals can learn to suppress both the cough and the corneal reflexes. Indeed, some reflexes can even be reversed. All of this makes it apparent that there will be no sharp demarcation between automatic and nonautomatic movements, and that it is probably best to accept Jackson's proposal that movements are on a continuum between least and most automatic. Furthermore, as discussed above, set and practice can shift the position of a given stimulus-response relationship on this continuum between the very automatic and the less automatic.

EFFECTS OF SET ON AUTOMATIC RESPONSES

A relatively automatic response that could be turned off or on by a prior instruction was observed by Traub et al. (1980) in an experiment in which the subject was asked to maintain thumb and index finger a few millimeters from the rim of a glass of sherry, as if about to pick it up. A muscle response (latency 50 to 60 msec) was evoked when the experimenter displaced the subject's wrist so that unless the subject grasped the glass it would be knocked over and broken. This short-latency grab occurred even though the displacement failed to stretch the muscles that were involved in the grab. The authors noted that while the "sherry-glass response" was automatic and could not be increased substantially by effort of will, it was nevertheless dependent on the subject's intent, for if he chose to ignore the fragility of the glass and knock it over, no response was observed. Here, then, is an example of a response that occurs "automatically," but for which the subject's motor set, as established by the experimenter's instruction, is critical. To what extent should this set-related, but rather short-latency, muscle activity be thought of as a reflex? This issue has been discussed by Traub et al. (1980) in connection with their observations on EMG responses to "indirect thumb displacement," a type of displacement in which the thumb itself is not directly perturbed, but in which the *wrist* is pulled to displace the thumb from the lever against which it is acting. It was observed that when flexor pollicis longus was indirectly unloaded by a pull on the wrist, EMG activity *increased* after 50 msec, rather than decreasing, as is usual when the muscle is unloaded and shortened directly. The response to the indirect unloading was not affected by anesthesia of the thumb, the skin of the wrist, or the motor nerve of the long thumb flexor. The authors therefore suggested that it was driven by a stimulus to proximal muscles or joints. Furthermore, the indirect response to unloading, which tended to move the thumb toward the lever, was elicited *only* when the wrist-pull removed the thumb from the lever, and disappeared immediately if the tracking apparatus was held in the hand. The authors therefore concluded that although the response occurred too early to be "voluntary," it nevertheless depended on the nature of the task and the subject's prior instruction. In our view, however, this sort of set-dependent response is the essence of volitional behavior and, though Traub et al. state that the response is too "early" to be volitional, we believe that the difference between the view that we express here and that of Traub et al. is really one of nomenclature rather than a conceptual distinction.

The experiments reviewed above have revealed that motor set interacts with peripheral inputs to generate outputs that are so fast as to seem automatic, *but that are nevertheless dependent on set*. In addition, as Traub et al. noted in the case of seemingly automatic and yet set-dependent responses in humans, the inputs eliciting them *need not arise from the muscles that exhibit the responses*, but may arise from totally different muscle groups. The nature of these set-dependent automatic responses, and the differences between them and the more familiar segmental reflexes, are sufficiently important to lead us to consider them in the motor control system for speech, which, like that for the hand, is highly represented in the cerebral cortex.

SET EFFECTS ON AFFERENT REGULATION OF SPEECH

Thus far we have concentrated on stretch reflexes that function in closed-loop negative feedback control systems. But afferent inputs may be utilized in open-loop as well as closed-loop systems. In what follows we cite several examples of short-latency motor acts that, while they might often be considered to be reflexes, are dependent on the motor goals of the subject: that is, they are volitional and set-dependent.

The term open-loop is sometimes used to imply a mode in which afferent feedback has no significance whatever, but this is not the correct way to use the term. The essence of a closed-loop negative feedback system can be appreciated from the classical description of muscle stretch reflexes: an increase in muscle length will cause greater discharge of muscle stretch receptors leading to muscle contraction and muscle shortening, which in turn will *reduce* the stretch receptor discharge. Thus, the output of the system tends to nullify the change in neural feedback that initiated the reflex response. In contrast, reflex effects in an open-loop system do not nullify the afferent input that initiated the reflex. An example of an open-loop system is the vestibulo-ocular reflex, in which a displacement of the head is signaled by the vestibular receptors and the resultant reflex is a compensatory movement of the eyes to maintain constant gaze orientation. The vestibulo-ocular reflex is open-loop in the sense that the reflex output (eye movement) has no effect on the discharge of vestibular nerve fibers that generated the reflex.

Abbs and Cole (1982) have provided a description of how open-loop systems control speech movements in which afferent signals originating from one moving part, the jaw, influence another moving part, the lips, and have proposed that many of the consequences of afferent inputs delivered in the course of speech depend upon open-loop control. To obtain evidence for open-loop control in multiarticular coordination, Abbs and Cole measured the effects of unanticipated loads applied to the jaw during coordinated movements of the upper lip, lower lip, and jaw during speech movements that required lip closure (e.g., saying "P"). They found that compensation for loads applied to the *jaw* depended on movements of the *upper and lower lips*. The "reflex" response to retardation of jaw closure involved a greater movement by the lips, and occurred within 50 msec. Abbs and Cole pointed out that "when the jaw contributed a large displacement to the opening or closure of the oral cavity, the upper lip and lower lip contributed proportionately less, and conversely." Thus, for control of speech, afferent pathways provide information used by open-loop systems, allowing the goals of motor behavior to be achieved in spite of unanticipated events. Furthermore, the response to a given jaw displacement depended on the sound that the subject intended to articulate, that is, on the subject's motor set.

Additional evidence for such mechanisms has been provided by Lubker and Gay (1982), who concluded that there is clear evidence for very rapid reorganization of movements toward articulatory goals, and that speakers are able to accomplish rapid, essentially instantaneous reorganization of vocal gestures in order to achieve these goals, even without auditory feedback.

Nashner et al. (1979, 1981) have demonstrated several striking instances of context- and goal-dependent postural reflexes. Their studies showed that reflexive EMG responses in muscles whose length was changed by a postural disturbance varied depending upon the requirements of the situation. Nashner et. al. (1979, p. 46) point out that "the resulting patterns of EMG activity were highly specific for each kind of displacement, and all subjects completely reorganized the pattern of activity from one form to another *within the first trials*, even immediately following unexpected stimulus changes" [italics ours].

We have now considered a number of short-latency, automatic, set-dependent responses that provide for goal acquisition. The properties of these responses have been elucidated in experiments involving disturbances injected into ongoing movements, but the true importance of these responses must be primarily in connection with normal, undisturbed, motor behavior: it is unusual to have one's articulatory movements stopped by external forces. In normal, unperturbed speech, the open-loop systems discussed by Abbs and Cole automatically take into account the status of the neuromuscular system with respect to the goal of the impending movement and generate an output that acquires the goal, even though the individual contributions of the numerous participating muscles are quite different, depending on a variety of normally operating factors. The importance of such automatic coordination of the different components that underlie undisturbed movements can hardly be overestimated.

Another clear example is respiratory-movement coordination during speech. Hunker and Abbs (1982) noted that speech requires coordination across the laryngeal, respiratory, orofacial musculature, and that the neural control process superimposed over these semi-independent motor systems would have to ensure their complementary contribution in achieving common goals. When the same goal is achieved in a number of different ways (as when we reach out to touch the same object with a series of different arm movements) there is said to be "motor equivalence" among those movements. Hunker and Abbs noted that the operation of motor equivalence implies that there are several levels of motor programming: a more general, higher level not implicated in control of detailed subcomponents, and a lower level that acts within the subcomponents—a concept similar to J. Hughings Jackson's higher and lowest level of motor organization. As an example, they cited the control of the respiratory system for speech, in which alveolar air pressure can be produced by many different combinations of recoil and active muscle forces, and a given lung volume can be produced by many different combinations of rib-cage and abdominal movements. They evaluated the automatic coordination of contributions to expiratory air flow by rib cage and abdomen; their data showed motor equivalence at several levels in respiratory motor control during speech, with rib cage and abdomen varying reciprocally in their common contributions to lung volumes. Furthermore, when initial lung volumes were unconstrained, they varied considerably, and such "spontaneous" variations, in turn, required reciprocal muscle adjustments to offset the varying recoil forces in achieving equivalent alveolar pressures. Even when expiratory lung-volume trajectories for a sentence were equivalent, the relative contributions of the rib cage and abdomen varied considerably throughout. Hunker and Abbs (1982, p. 946) concluded that:

> . . . motor equivalence coordination permits, with apparent ease, substantial flexibility in the detailed patterns producing these complex voluntary behaviors.

We could, of course give further examples of set-dependent automatic responses. Our aim, however, is to provide a background for a neurophysiological analysis, rather than to provide a comprehensive review. We feel that the data presented are sufficient to show that set can modify the EMG at latencies of 50–70 msec and that set-dependent EMG responses show features that are commonly associated with reflexes, but at the same time have some properties of volitional movements. Further, an animal moving in a natural environment is capable of making extraordinarily complex open-loop adjustments to achieve its goals, overcoming a number of unpredictable obstacles, as well as the vagaries of the skeletomotor system itself. The latency for the effects described in this chapter, 50–70 msec, is easily compatible with the idea, put forward here and by many others elsewhere, that the cerebral cortex provides a mechanism for this dynamic function of the brain.

Chapter 4

Cerebral Cortex: Generalities and Specializations

BEHAVIORAL FLEXIBILITY AND THE CEREBRAL CORTEX

It is commonly believed that the mechanisms underlying set-related phenomena and behavioral flexibility, as instanced by rapidly changeable input-output relations, lie within the cortex, although the evidence for this proposition is less clear than one might expect. As Phillips (1981, p. 151) states in reference to the enlarged cortex of humans relative to nonprimates:

> Big brains have given to the primates the general advantages for survival that attach to adaptive behaviour which is less "stimulus-bound" than that of lower animals: exploratory drives; internal trial-and-error; orientation to what may be happening beyond the range of sight, smell and hearing; and knowledge of the consequences of particular actions, as in copulation, sowing and harvesting.

The neocortex is, of course, the largest and phylogenetically the most recently developed part of the cerebral cortex. It exists, in a clearly recognizable form, only in mammals and is especially well developed in primates. However, it is not solely in humans and other primates that the cortex has increased dramatically in relation to body size over the course of evolutionary development. Disproportionate increases in brain size have also occurred in many other groups of mammals. These increases in size are not without cost (see, e.g., Armstrong, 1983), so they must yield substantial benefits. In humans, the neocortex probably accounts for as much as 15% of the body's oxygen consumption, although it constitutes less than 1.5% of its weight. This massive metabolic expenditure alone demonstrates the importance of the neocortex. Still there is a noteworthy lack of a succinct statement of the function of the neocortex. Barlow (1978, p. 79) expressed this view particularly well in asking:

What does the cortex do? As a physiologist I am strongly prejudiced to believe that, for an organ of such uniform structure, there must be a simple answer to this question. The heart pumps blood; the lungs exchange gases; the kidney forms urine; and the retina transduces the optical image and transmits it to the brain. It must be worth searching for the appropriate succinct description of what the cortex does.

It seems possible, in this context, that the neocortex serves to do what mammals, in general, do: flexibly interact with changing patterns of environmental events.

It is usually held that large brain size, which in mammals is due mainly to the size of the neocortex (Jerison, 1977), is correlated with arboreal locomotion (Eisenberg and Wilson, 1981), and with a need, during the life of a long-lived individual, to acquire information from experience to guide its behavior (Eisenberg, 1981), and to gather food from dispersed regions (Eisenberg and Wilson, 1978; Clutton-Brock and Harvey, 1980). While these observations are no more than intriguing correlations that imply nothing about causality, they point to an assembly of features related to increased behavioral flexibility.

However one views the relationship between brain (and neocortex) size, fitness, and intelligence, one thing seems clear: mammals, as a group, can be characterized by their flexible interaction with the changing external environment (Warren, 1974). The central feature of animal intelligence is not ability at simple learning, but rather at higher-order learning with reversal of cues or capacities to adapt flexible strategies (see, for example Jerison, 1977; Passingham, 1981). If this is so, then the neocortex would appear to be the most likely candidate for a neural structure that subserves such flexibility.

Ethologists have viewed the behavior of mammals as "less stereotyped, more individually variable, and more strongly influenced by learning than the behavior of other vertebrates." According to Ewer (1968, p. 349), "The most striking behavioral characteristic of mammals is their learning ability. Compared with other animals, mammals can learn faster, learn more, remember more and show more insight." In parallel, neuroanatomists have long known that the most fundamental difference between the brains of mammals and other vertebrates is the specifically mammalian neocortex. This correlation between mammalian structure (neocortex) and function (behavioral flexibility) suggests that the cortex may be necessary for the behavioral flexibility that characterizes the mammals as a group. However intuitively satisfying that formulation may be, we must admit that we know of little direct scientific information that supports this view (see Kolb and Whishaw, 1981; Whishaw et al., 1981).

We hasten to point out that one cannot view behavioral flexibility as the sole province of the neocortex. Spinal-cord circuitry also demonstrates a measure of flexibility even when disconnected from the brain. Following the line of Sherrington's initial studies (see Chapter 2), Forssberg and his colleagues (Forssberg, 1979; Forssberg et al., 1980a, b) have studied "reflex reversal" in a locomoting cat. A cutaneous stimulus can elicit a limb flexion during the phase of movement characterized by swinging the limb forward and a limb extension when the limb is approaching a surface and is about to support the weight of the body. However, the capacity of the spinal cord (or other subcortical structures) to produce such flexibility should not lead to the conclusion that the cortex plays no role in these functions. Elaboration of something like the spinal reflex reversal mechanism at the cortical level may give rise to the flexibility characteristic of mammals in their interaction with the environment.

ORGANIZATION OF THE CORTEX

If the cortex is so important to the higher brain functions we wish to investigate, we have to learn more about its structure and organization. Fortunately, although much work needs to be done, a substantial amount of progress has been made on this problem in the past decade, largely as a result of improvements in neuroanatomical methods. What follows is a brief account of the organization of the neocortex. Some of the issues mentioned here are dealt with in more detail in Chapter 5, but it is worthwhile to review the basic features of cortical functional organization for two reasons: first, because these principles are subject to certain misunderstandings; and second, because it allows us to introduce the nomenclature used in this monograph. The inconsistent and bewildering thickets of nomenclature in the literature make it almost impossible for the nonspecialist to follow many recent developments in the field.

The cortical sheet, however folded, can be thought of as having two dimensions: horizontal and vertical. The vertical organization is laminar; the cerebral cortex is considered to have six fundamental layers. Each layer has a characteristic set of neural inputs, output targets, and intrinsic connections. For our purposes, the most important observation is that the output of the cortex to the brainstem and spinal cord originates exclusively from cell bodies located in one cortical lamina, layer V. The output of one part of the cortex to another originates from many cortical layers, but in many cases the contribution from cell bodies in layer III is largest. These corticocortical neurons send axons which terminate in many layers, although they also most commonly terminate in layers relatively near the cortical surface (layers II, III, IV).

The horizontal organization is somewhat more complicated. The cortex can be divided into a number of functionally and (usually) structurally distinct areas termed cortical fields. There are two somewhat contradictory schemas by which these fields are grouped. In one view, the cortex consists of a few primary and secondary motor and sensory areas, while the vast expanse of cortex is thought of as an "association cortex" (see, e.g., Eccles, 1981). What is being "associated" is not usually made very clear, but these areas are sometimes thought to be polymodal in some sense. Against that view is the concept that most of the cortex lies within the domain of one of the sensory modalities (see, e.g., Diamond, 1979) and that there may be a few, restricted polymodal areas. The major sensory domains are the visual, auditory, and somatosensory, to which a motor domain and, perhaps, a "limbic" domain might be added. The somatosensory areas are often linked with the motor areas of cortex (Diamond, 1979; see also Woolsey, 1958) as a somatic sensorimotor domain.

From the beginning of the modern era, the fields of the cerebral cortex have been defined on the basis of a constellation of properties (Rose and Woolsey, 1949). Currently, the properties that have received the most attention include: cortical architecture, patterns of neuronal connectivity, properties of neurons during behavior or after experimental manipulation, the effect of artificially induced electrical activity, the behavioral consequences of cortical damage, and the distribution of neuromodulatory agents and their receptors in the cortex. Although the analysis is by no means complete, it is now clear that certain areas are more closely linked with the skeletal motor-output systems we will focus upon. This general region, which is comprised of a number of cortical fields, is termed the *somatic sensorimotor cortex*. In monkeys, it consists of a large number of cortical fields, many of which are *somatotopically organized* and contain a complete (or nearly complete) *representation*, or map, of the body. There is some evidence that each field may play a specialized role in the cerebral control of movement.

The concept of multiple representations of a sensory modality or the skeletal motor apparatus is now generally accepted for the cortex as a whole, and dates at least from the era of J. Hughlings Jackson (1931). The many visual and auditory cortical fields have been well documented and the demonstration of multiple somatosensory cortical fields is also gaining acceptance. However, comparable knowledge concerning the motor areas of cortex has proved to be something of a stumbling block. Although the total extent and the exact number of fields of the somatic sensorimotor cortex are not yet known, recent investigations, largely with microelectrode methods, have greatly improved the understanding of this problem.

ORGANIZATION OF THE CORTEX

In examining the question of how many cortical fields can be identified within the somatic sensorimotor cortex, one inevitably begins with the classic work of Woolsey (1958; Woolsey et al., 1952) in nonhumans and of Penfield (Penfield and Rasmussen, 1952) in man. Woolsey defined cortical regions as "motor" on the basis of the effects of surface stimulation, and divided somatic sensorimotor cortex into four subdivisions (Figure 4.1): rostral to the central sulcus was the precentral motor cortex (MI); caudal to that sulcus was the postcentral somatosensory cortex (SI); rostrally on the medial surface of the hemisphere was the supplementary motor cortex (MII); and laterally in the lateral sulcus was the second somatosensory cortex (SII). However, it is now known that Woolsey's estimate of the number of cortical fields in the somatic sensorimotor cortex was conservative. There may be about a dozen fields within this domain, although the precise number has not yet been determined for any species.

On purely neuroanatomical grounds (Jones and Powell, 1969), SI has long been thought to contain several separate and structurally distinct cortical fields. This proposition, though still not without controversy (cf. McKenna et al., 1982), has been confirmed with neurophysiological methods within the past few years. Kaas et al. (1981) have shown that there are at least two, and probably four, separate representations within SI, each corresponding to one of the structurally defined fields and each with characteristic responses to sensory stimulation.

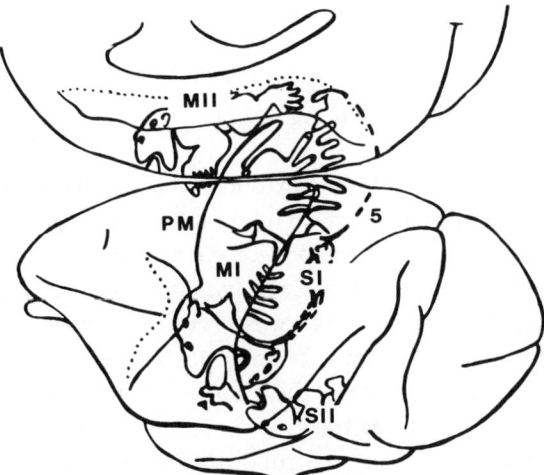

Figure 4.1. *Somatic sensorimotor cortical fields.* Surface view of the left hemisphere of a macaque monkey brain. The four somatotopically organized regions identified by Woolsey and his colleagues are indicated by the figurines and are meant only as the most general guide to somatotopy. Rostral is to the left. Abbreviations: MI, precentral motor cortex; MII, supplementary motor cortex; PM, premotor cortex; SI, postcentral somatosensory cortex; SII, second somatosensory cortex; 5, area 5 of the posterior parietal cortex. (After Woolsey, 1964)

Within MI, as originally described by Woolsey, it has been found that there are regional specializations which, while not yet clearly satisfying all of the requirements for consideration as separate cortical fields or representations, indicate that significant functional divisions exist. The vast majority of MI neurons, activated by stimulation of hair or skin receptors, are concentrated within a subdivision of MI closest to the central sulcus (Strick and Preston, 1982). Furthermore, stimulation of MI with microelectrodes yields a pattern of movements that has been interpreted as indicative of multiple motor representations (Strick and Preston, 1982).

Similar microelectrode studies within and near the region defined as the second somatic sensorimotor cortex (SII) have revealed at least two, and perhaps three, representations, each corresponding to a structurally defined field with a unique set of response characteristics (Robinson and Burton, 1980). Medially, the MII cortex seems to be much as Woolsey described it, although somewhat more rostrally situated. Rostral to MI, however, is a nebulous zone that has often been ignored in the past. This is the premotor cortex (PM), so-called because it lies in front of the precentral motor cortex (MI). Recent efforts to define this area have yielded a consensus that PM is a motor field separate from MI with a characteristic cytoarchitecture and connectivity, among other features (Wiesendanger, 1981; Humphrey, 1979b; Wise, 1983). The posterior parietal cortex, area 5 caudal to SI, may also include more than one cortical area. There is evidence for a medial area, termed the supplementary sensory cortex (Woolsey et al., 1979), as well as the more extensively studied lateral area on the superior parietal lobule.

CEREBRAL CONTROL OF MOVEMENT

If the cortex is essential for the flexibility of stimulus-response relations, what, then, is the role of each cortical field in the cerebral cortex in movement? Until the areas of the somatic sensorimotor cortex are defined more precisely and until much more physiological study is done on these fields, very little can be said about the different motor roles of each individual cortical field. However, some information has been collected concerning their function as a group, since they all send their corticospinal and corticomedullary fibers through the pyramidal tract.

The most profound effect of destruction of the pyramidal tract (Lawrence and Kuypers, 1968) is that monkeys so damaged never recover the ability to make the fine, independent finger movements of the sort necessary to remove objects from shallow wells by digital manipulation, or to move their limbs with normal speed and agility. On these grounds it has been concluded that the pyramidal tract system in monkeys is important for adding an extra degree of motor flexibility to the motor system as a whole. Lawrence and Kuypers (1968, pp. 9–10) asserted that:

> The effect of bilateral pyramidal tract interruption is most clearly demonstrated in the eight animals in which the lesion was virtually limited to the pyramidal tracts. In these animals, following operation, there was an immediate ability to sit with the head up and to stand, walk, run and climb. In addition, they were able to cage place and chin place. Yet during this early post-operative phase they were apparently unable to use their extremities, especially their hands, independently of total body movements. Thus, although they could not pick up food with their hands they could use them in clinging to the cages and in climbing. After further recovery, however, they regained the capacity for independent use of their extremities and within three weeks could reach accurately with either hand to pick up morsels of food by closure of all fingers in concert. The reaching movement initially consisted of a hooking circumduction of the whole arm, with closure of the hand being part of the total arm movement. During further recovery, closure of the hand became progressively more independent of the arm movements. Ultimately the animals could fully extend either arm with the wrist slightly dorsiflexed and the fingers semiflexed and abducted. With the arm in this stable position, they were able to open and close the hands to pick up pieces of food. Yet in spite of this recovery, individual finger movements never returned, even after recovery periods of up to eleven months. In addition, all movements were slower and fatigued more rapidly than in the normal animal. Finally, there was a persistent difficulty in releasing the grip after picking up food, although in climbing it appeared to be released readily. The sequence of recovery in the affected limbs following unilateral pyramidal tract interruption was similar to that in animals with bilateral lesions, but in order to force the maximum return of function in the affected arm, the unaffected arm had to be periodically restrained.

A few words of warning should be noted about these conclusions. Whereas many authors have taken the findings of Lawrence and Kuypers to imply that the cortex controls *only* "relatively independent" movement of the fingers, there is strong evidence that whatever the motor cortex does, it does for the body as a whole. Rather than emphasizing the role of cortex specifically in hand movements, which are particularly important in humans and many other primates, it is now generally accepted that, at least in monkeys, one of the roles of the cerebral cortex in movement is to enable the animal to make fractionated voluntary movements of any motor organ. As Lawrence and Hopkins (1976, p. 246) warned:

> . . . it must be emphasized that [relatively independent finger movements] are merely a convenient representative of a wide range of normally occurring discrete movements. Such movements are most abundant in the arms and

legs, particularly distally, but they also occur in the trunk. All of these movements, in addition to RIFM [relatively independent finger movements], are abolished by pyramidal [tract] lesions.

Hence, it would appear that if the roles of cortical and subcortical structures in the control of movement differ, the specific role of the cortex in voluntary movement lies in a supplementation of the activity of the subcortical motor system, particularly with respect to improvements in the flexible and fine use of the motor apparatus.

FUNCTIONAL SPECIALIZATION OF THE PRECENTRAL MOTOR CORTEX

Many observations point to the precentral (or primary) motor cortex as being the main cortical field necessary for this flexibility (Evarts, 1981). We will mention only two here:

1. Most of the neurons within this field "saturate" with small movements. That is, they show their full range of firing-frequency modulation during and before small movements. Thus, the idea that the MI is specialized for the control of fine, accurate movements draws support from physiological and behavioral studies.
2. Corticospinal neurons in precentral motor cortex show more modulation in their discharge pattern when monkeys make fine, independent finger movements than when they make coarse, powerful finger movements (Muir and Lemon, 1983).

An important question concerning the functional organization of the motor cortex is that of "muscle versus muscles": Do individual cortical neurons specify activity in each muscle separately? A related question is that of "muscles versus movements": Do these neurons control muscle activity or specify instead the movements that are made per se? The answers, although still the subject of strenuous debate, now seem fairly clear to us. First, cortical neurons do not appear to control each muscle separately. One of us (Shinoda et al., 1981) has shown that single neuronal fibers originating in the MI cortex branch in the spinal cord to innervate the motoneurons of a number of different muscles. As far as is known, all of these connections are excitatory. Thus, the cortex influences related muscles together, presumably in a coordinated manner. Similarly, when one examines certain MI neurons, the activity of which is well correlated with muscle activity (see Chapter 10, and Cheney and Fetz, 1980), one finds that the activity of these cells may be correlated with the activity of several muscles. As for the second question, the cortex seems to influence muscles—that is, muscle activity—rather than movements per se. In the clearest example, the activity of almost every neuron in the MI cortex that is directly correlated with movement from one position to another is also well correlated with force generated by the same muscles (Evarts et al., 1983). (For a review of these issues see Evarts, 1981.)

How input to these cells might be processed to achieve the observed flexibility of the stimulus-response system and how this input-output information-processing might be studied, forms the topic of the remainder of this monograph.

Chapter 5

Cell Types and Information-Processing Circuits in the Cerebral Cortex

It now appears, on the basis of the research of the last 25 years or so, that the neocortex has a modular form of basic organization that allows it to process incoming information both serially and in parallel, and to direct that information to a number of at least partially independent outputs. We use the term "information" here in the broadest sense, to mean all of the inputs to the cortex. In this chapter, we discuss some of the organizational features of the cortex and how these features might form the basis of the neural switching processes we will discuss later (Chapters 8, 9, and 10).

COLUMNAR ORGANIZATION

As mentioned above, one of the key concepts of cortical input-output information processing is that of columnar organization. The principles of columnar organization were summarized by Mountcastle (1978, p. 16):

> The idea of the columnar organization of the cortex has developed as a functional concept on the basis of a discovery made in physiological experiments, namely that the basic unit of operation in the neocortex is a vertically arrayed group of cells heavily interconnected along that vertical axis, sparsely so horizontally. This unit is envisaged to function in the operations of processing and distribution. . . .
>
> 1. The cortical column is an input-output processing device. The number of other regions transmitting to and receiving from a traditionally defined cortical area may vary from about 10 to 30. The sample of that total entertained by any given subset of modules of an area is much smaller and varies among subsets, with overlap.
> 2. The columnar arrangement allows the mapping of several variables simultaneously in a two-dimensional matrix.

3. Specific connections are maintained between ordered sets of columns in different cortical areas and between sets of cortical columns and modules of subcortical structures. Thus, topological relations may be preserved during transit through and between such areas, with or without topographic (geographic) mapping.
4. The identification parameters for columns and ordered sets of columns may vary within a given cortical region, as defined traditionally, and may differ strikingly between cortical regions.
5. The columnar functional model allows for a partially shifted overlap across a topographical representation that is compatible with a dynamic isolation of the active elements of a column by a form of lateral, pericolumnar, inhibition.
6. Divergent intracolumnar pathways to different outputs allow selective processing ("feature extraction") of certain input signal parameters for particular output destinations.

The major points, for our purposes, are numbers 1 and 6. We seek to determine the relative strengths of inputs to a cortical column and how these might vary as a function of time. Also, it is important to determine the extent to which the different output pathways from a given cortical column can carry distinct information.

INPUTS TO CORTEX

One might suppose the inputs to the cortex to be of three types: relays of information or signals into a cortical field; *diffuse* modulators of large brain regions that have a uniform effect on input-output processing; and *specific* modulators, or gating agents, that selectively potentiate specific input pathways to a cortical field and thereby accentuate their influence over cortical output. Substantial progress has been made in recent years in understanding certain of the inputs to the cortex, in the context of these ideas.

Specific Thalamocortical Inputs

In the classical view, the specific thalamocortical inputs were believed to terminate almost exclusively on nonpyramidal (stellate) cells in the fourth cortical layer, whence the information was thought to be transferred to deep layers for further processing and output (Lorente de Nó, 1949). In the view of Lorente De Nó (p. 305):

> From the functional point of view it [the cortext] is a unitary system composed of vertical chains of neurons, among which anatomically the most important

are those starting at the articulation of the specific afferents and the cells of the external lamina [layers II, III, and IV]. The neurons with descending axons of this lamina send their impulses to layers V and VI, from which impulses are sent back chiefly to layers II and III. . . .

There remains some heuristic merit to this concept, but the actual situation is now known to be more complex. It has been determined, at least in the mouse first somatic sensory cortex (SI) and the rat visual cortex, that, in addition to thalamic inputs to dendrites of nonpyramidal cells in layer IV, thalamocortical terminals are found on virtually all dendritic elements in that layer (see White, 1981). These elements include those from superficial pyramidal cells with basilar dendrites in layer IV, and especially from deep pyramidal cells, which send apical dendrites through layer IV (Figure 5.1). In many cortical fields, layer III pyramidal cells also send dendrites into layer IV. If White's finding is a general one, common to all species and cortical areas, then the opportunity for elaborate parallel processing networks is large.

An important example of parallel processing is found in the visual system. It was originally proposed (Hubel and Wiesel, 1962) that inputs from the

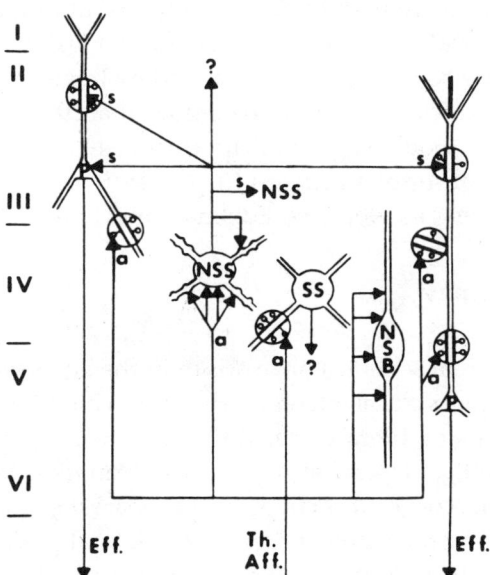

Figure 5.1. *Cortical cell types that receive direct thalamic input.* Abbreviations: Eff., cortical efferent; Th. Aff., thalamic afferent; SS, spiny stellate cell; NSS, nonspiny stellate cell; NSB, nonspiny bipolar cell; P, pyramidal cell; I–VI, layers of the cerebral cortex; a, asymmetrical contacts; s, symmetrical synaptic junctions. (From White, 1981, figure 2)

visual relay nucleus, the lateral geniculate nucleus, to cortex would be only to cells with simple receptive fields. These "simple cells" then would converge upon other cortical cells to form the various types of complex fields. It is now fairly well accepted, however, that in addition to this *serial information processing*, there is a parallel input to visual cortex neurons, e.g., lateral geniculate neurons project to many cell types within visual cortex. Toyama et al. (1981a,b), who used cross-correlation techniques, and Bullier and Henry (1979a,b,c), who used more conventional analytical methods, concluded that parallel processing of visual signals is a prominent feature of intracortical information processing. Bullier and Henry have reported substantial parallel processing in cortex of inputs from X and Y retinal ganglion cells. Toyama et al. concluded that many complex cells receive thalamocortical inputs as well as inputs from local circuit neurons. This conclusion concurs with other studies which showed that striate cortex cells with both complex and simple receptive fields receive direct inputs from lateral geniculate cells (Stone and Dreher, 1973; Singer et al., 1975). Intracortical mechanisms, including intrinsic inhibitory GABAergic cells (Sillito, 1975, 1979), appear to shape the specific and distinct visual receptive properties of striate cortex neurons receiving this parallel input. Thus, we can conclude that both serial and parallel information processing are probably important to striate cortex function, and are both likely to be important to the function of other cortical fields, as well. Essentially the same conclusions have been drawn from studies of the somatic sensorimotor cortex. In general, the anatomically described inputs to cortex seem to be cruder than the responses of cells in these cortical fields would suggest. It appears likely that inhibitory "shaping" or "sharpening" of input signals to the cortex is important in the somatic sensorimotor cortex, as well (see Edelman and Finkel, 1983).

Corticocortical Inputs

The same conceptual schemes have arisen in the analyses of another major cortical input, corticocortical afferents. Here, as with intracortical information processing, serial (or hierarchical) information processing has been the most well studied, but parallel routes of information flow are also likely to be important in cortical function. In the clearest example, it is held that visual information enters the cortex via the lateral geniculate input to the striate cortex. The input is then processed within the striate cortex according to the scheme outlined above. Subsequently, some of the output neurons send this information to the peristriate fields in much the same way that the lateral geniculate nucleus sends input to the striate cortex. This peristriate cortical field then processes the information and serially transfers it to another cortical field, and so forth (Rocha-Miranda et al., 1974). Anatomical connections appear to be consistent with this idea (Jones and Powell, 1970; Pandya and Kuypers, 1969), as are studies which attempt to disrupt visual function by disconnecting one cortical field from another by surgical intervention or reversible coding lesions (Schiller and Malpeli, 1977). Further, anatomical information has suggested laminar asymmetry in these corticocortical connections. Those cell bodies that project from areas "closer" to the primary visual area, the striate cortex, to fields "further" from the striate cortex in this hierarchical scheme, are found mainly in superficial layers. These corticocortical fibers terminate in the granular and superficial layers (much like the specific thalamocortical projections). The corticocortical projection in the "reverse" direction is said to arise predominantly from deeper layers, mainly layer VI (much as the cells in striate cortex that project back to their thalamic input, the lateral geniculate nucleus, arise from layer VI), and terminate in the outermost layer of cortex, the neuron-sparse molecular layer (Rockland and Pandya, 1979; Tigges et al., 1981).

Although this hierarchical model of corticocortical function has some convincing evidence behind it (e.g., Schiller and Malpeli, 1977), there are sound reasons to believe that the situation is not so simple and that other processes act as well. To consider these other processes and pathways, we must digress to a consideration of thalamocortical pathways. Neuroanatomically, it is clear that visual information can reach extrastriate (nonprimary) visual areas through routes not including either the lateral geniculate nucleus or the striate cortex. To cite one well-documented example, the retinal projection to the superior colliculus is continued by a collicular projection to part of the thalamus, the LP-pulvinar complex, which projects to extrastriate cortex. As Graybiel and Berson (1981, p. 286) have summarized on the basis of their own and earlier studies of cat visual cortex:

1. Recent electrophysiological mapping studies of the visual cortex have demonstrated the presence of as many as thirteen retinotopically ordered areas in the posterior association cortex (Allman and Kaas, 1975; Palmer et al., 1978; Tusa et al., 1975; Van Essen, 1979; Zeki, 1978).
2. The transcortical association pathways leading out from the primary visual cortex terminate in some but not all of these extrastriate visual areas (Van Essen, 1979; Zeki, 1978).
3. There are, in addition to these transcortical routes, multiple lines of ascending conduction that reach the extrastriate cortex by way of transthalamic pathways synapsing in the nucleus lateralis posterior-pulvinar (LP-pulvinar) complex (Berson and Graybiel, 1978; Glendenning et al., 1975; Graybiel, 1972).

Graybiel and Berson (1981, p. 286) conclude that: "the visual cortex is not organized according to a single striate-to-extrastriate hierarchy but according to a more intricate plan that takes into account, as one determinant of transcortical connectivity, the thalamic affiliations of each cortical zone." Further, although destruction of the striate cortex has devastating behavioral effects, residual visual function can be documented (as first shown by Humphrey, 1970, and Pasik and Pasik, 1971). In accord with this view, inactivation of the cat's striate cortex by cooling it below the temperature at which synapses will function or axons will conduct action potentials has

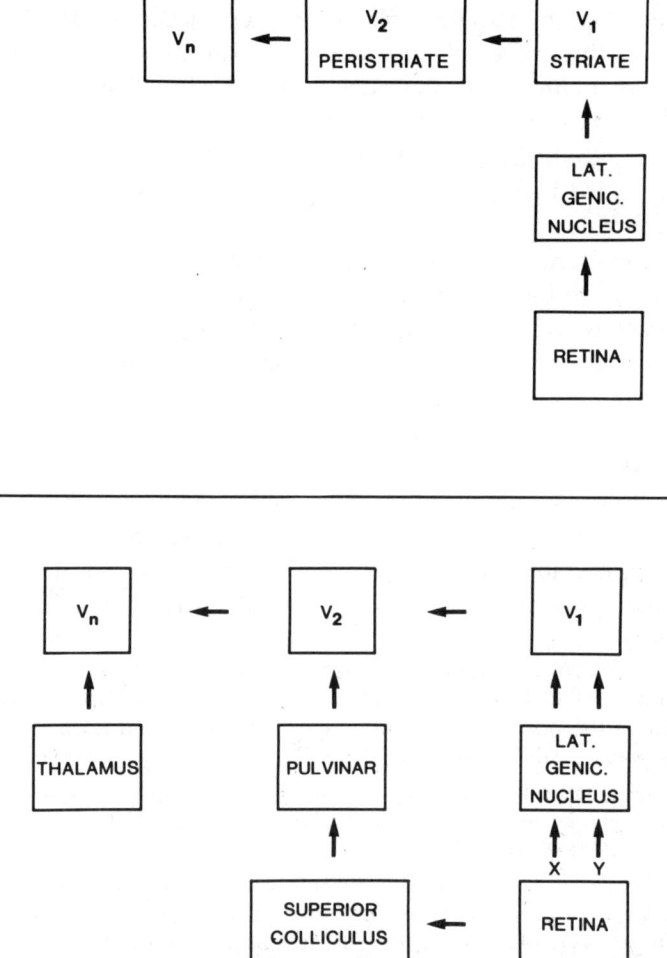

Figure 5.2. *Contrasts between serial and parallel processing* are depicted in this highly simplified diagram of serial visual information processing (above) and parallel processing (below).

little or no effect on the visual receptive field properties of peristriate cortex cells (Sherk, 1978). It is apparent that there are important parallel routes by which visual information reaches certain areas of cortex and which do not always depend upon cortical fibers (Figure 5.2). These parallel routes may vary in functional significance for different fields, species, and sensory systems, but seem to be uniformly available. We hasten to point out that the organization of visual corrections is not, in any species, as simple as that illustrated in Figure 5.2. For example, recent studies in monkeys have shown that the primary visual field, the striate cortex, receives an input not only from the lateral geniculate nucleus, but also from the LP-pulvinar complex (Benevento and Yoshida, 1981; Miller et al., 1980).

In addition to the various thalamic inputs, information to cortical fields could arise from a variety of cortical areas in both hemispheres, not just one hierarchically "lower" field in a sensory progression. To some extent, it appears that these inputs from the converging cortical sources terminate in anatomically distinct parts of that cortical field. The extent of this segregation seems to be variable. In some areas and species, for example frontal cortex of the macaque monkey, it appears that commissural and ipsilateral corticocortical inputs are at least partially segregated in different parts of a target cortical field in an interdigitating manner (Goldman-Rakic and Schwartz, 1982). In other areas, such as extrastriate visual cortex in the macaque, there appears to be exact correspondence between ipsilateral and contralateral cortical inputs (Van Essen et al., 1982). In still other areas the situation is ambiguous (Jones et al., 1979a; Goldman-Rakic and Schwartz, 1982), with some parts of an ipsilateral projection in register with commissural input and other parts alternating. The existence of overlapping corticocortical termination leaves open the possibility of tremendous convergence of projections from a variety of cortical fields onto individual neuronal elements or circuits. Such convergence of parallel thalamic and cortical inputs could serve as the "raw material" for set-dependent changes in input-output coupling at the cellular level (see Chapter 8).

Modulatory Inputs

It can be imagined that subcortical inputs could provide a signal that is selectively potentiated (or inhibited) by other afferent pathways to the cortex. We develop this view in detail in Chapter 8. For the present, it is worth noting that the concept of modulatory inputs to the cortex—that is, inputs that modify incoming signals—is not new. Jasper, Magoun, and Lorente de Nó postulated on separate grounds the existence of such modulatory inputs to cortex, such as may occur during different states of arousal. Usually such modulating inputs have been postulated to be diffuse or unspecific (Lorente de Nó, 1949) and to arise from the intralaminar thalamic nuclei. It has been found that certain other thalamic nuclei in cats and rats indeed project rather diffusely to the cortex, that is, to a broad extent of cortical territory (Herkenham, 1980; Glenn et al., 1982). However, while their projection to a wide area of cortex does not necessarily imply that these thalamic nuclei are nonspecific in terms of the neural elements they influence in the cortex, these diffusely projecting nuclei may play a role in gating at the cortical level (see Chapter 8). Other potential sources of modulation at the cortical level include the monoaminergic inputs from the locus coeruleus and the raphe nuclei. These projections have also, in the past, been considered diffuse or even homogeneous in their distribution. However, recent analyses have shown that, in the monkey, these fiber systems are distributed to the striate cortex in restricted laminar patterns that differ markedly for the noradrenergic and serotonergic inputs (Morrison et al., 1982), and preliminary evidence suggests that these systems play a role in the responses of cells to sensory stimulation (Waterhouse and Woodward, 1980; Kasamatsu and Heggelund, 1982). Other subcortical inputs, for example the basal nucleus of Meynert (Jones et al., 1976; Kievet and Kuypers, 1975), claustrum, and amygdala (Krettek and Price, 1977), remain relatively unstudied in this regard, but may also serve as modulators of input pathways. Thus, there appear to be a number of candidates for pathways involved in either specific or diffuse modulation of inputs to a cortical column.

CORTICAL OUTPUT ORGANIZATION

A major question that arises concerning the cortical output system is: Does a column of cortex have a single, unified output to its various targets, or is its output fractionated in such a manner that it could direct distinct information to different targets? The answer seems to lie between the two extremes. Clearly, projection neuron systems of the cortex have the capacity to fractionate output to a substantial extent, an extent sufficient to require that the functional roles of the different corticofugal cell types be considered separately. There is nothing inherent in the idea of columnar organization that requires all cells in a column to be doing the same thing at the same time. Indeed, differences are to be expected. We will consider three different combinations of output systems in an effort to determine their independence from each other:

1. corticocortical versus corticofugal,
2. epsilateral versus contralateral corticocortical,
3. different corticofugal projections.

The corticocortical and commissural cells, at least those in the superficial layers, can send outputs that are completely different than those from the corticofugal cells, which are mainly, if not exclusively, in layers V and VI (Table 5.1). More generally, it may be that the descending corticofugal outputs and corticocortical outputs also are largely separate even in layers where they co-exist, such as layer V. For example, Catsman-Berrevoets et al. (1980) have shown that in the rat somatic sensorimotor cortex, no corticospinal cells send axon collaterals to the contralateral hemisphere. Although more data are needed, the output from a cortical field to other cortical areas appears to be largely separate from its output to subcortical targets. It must be noted, however, that this distinction is not absolute; evidence has been obtained for the existence of at least some cells that project from layer V of the somatic sensorimotor cortex of cat to both another cortical field and to brainstem or spinal structures (Deschenes, 1977; Zarzecki et al., 1978a).

The ability of different corticocortical cells in a given cortical field to send separate signals to their various cortical targets is conceivably more limited, but seems likely to be extensive on the basis of recent findings. There is now some evidence that commissural and ipsilateral corticocortical fibers, although their cells of origin overlap greatly in laminar extent, are

Table 5.1. Laminar Distribution of Somata of Cortical Efferent Cell Types in Monkey Somatic Sensorimotor Cortex

Layer I		
Layer II	Short corticocortical	
Layer III	Longer corticocortical Callosal	
Layer IV		
Layer V	(Some long corticocortical)	Corticostriatal Corticorubral Corticothalamic (intralaminar) Corticobulbar Corticospinal and corticotectal
Layer VI	(Some callosal-motor cortex)	Corticoclaustral Corticothalamic (relay nucleus)

Source: After Jones, 1981a.

completely separate output systems in the rabbit visual cortex (Swadlow and Weyand, 1981) and are largely separate in the monkey frontal and parietal cortex (Schwartz and Goldman-Rakic, 1982; Anderson et al., 1982).

The degree to which the corticofugal output system can send different signals to its various targets also is not completely resolved. For example, corticothalamic axons to main relay nuclei arise from layer VI, whereas other corticofugal cell bodies (for example, corticospinal, corticotectal, and corticopontine) are situated in layer V. It is now certain that some fibers branch to send axons to more than one target structure (Donoghue and Kitai, 1981; Rustioni and Hayes, 1981; Swadlow and Weyand, 1981; Catsman-Berrevoets and Kuypers, 1981; Baker et al., 1983; Raczkowski and Diamond, 1978). One of many examples is a layer V cell in rat somatic sensorimotor cortex that sends axons through the cerebral peduncle to terminate in the brainstem or spinal cord, as well as sending branches to the striatum (Donoghue and Kitai, 1981). But it appears that only a minority of the output cells projecting from the cortex to one structure sends axons to any other structure. For example, only one-third of corticotectal cells in rabbit visual cortex send a collateral to the thalamus (Swadlow and Weyand, 1981), and only about one-tenth of the cells that project from the monkey's motor cortex to the red nucleus send collaterals to the spinal cord (Humphrey and Corrie, 1980). Therefore, there appear to be several independent pathways out of a cortical column, and great potential exists for selective transfer of cortical information to various cortical and subcortical targets. This is not to say that collateral branching does not occur in the corticofugal system or that it is unimportant. For example, with one output cell type—the corticospinal neuron—a number of investigators, using a variety of techniques (Shinoda et al., 1979, 1981; Futami et al., 1979; Fetz and Cheney, 1980; Cheney and Fetz, 1980) have demonstrated that individual corticospinal neurons branch to innervate *several* motor nuclei: occasionally the fibers from a single cell end in both forelimb and hindlimb parts of the spinal cord.

NEOCORTICAL CELL TYPES

Before approaching the subject of intrinsic cortical organization, one must consider the concept of cortical cell types. By a "cell type" we mean a neuron of at least partially known morphology and connections. We use the term here to refer to cortical neurons whose axonal targets have been at least partially identified. Of course, further knowledge about the functional organization of the cerebral cortex may ultimately lead to the further subdivision of each of the cell types that we can now single out on the basis of morphology and/or connections. However, the present definition serves as a starting point. Our use of the term cell type is different from the term as applied to invertebrates, where it is possible to identify an individual neuron that has similar structure and function in all individuals in a given species. In the mammalian cortex, on the other hand, it is impossible to identify an individual cell in one animal and then locate the same cell in a second animal. However, a variety of techniques applicable to the cortex allow a substantial amount of information to be gained about a population of cells whose activity is monitored extracellularly. For instance, one can locate a population of a functionally and structurally defined cell in one animal and then proceed to locate a corresponding population in another animal. By gaining a knowledge of a cortical neuron's projection, or at least a part of its projection, a great deal of correlated information becomes available. For example, neurons that project to the brainstem, spinal cord, and from one cortical field to another almost invariably have pyramidal-shaped somata, an oriented system of basilar dendrites, and apical dendrites that extend to near the surface of the cortex.

INTRINSIC CORTICAL ORGANIZATION

Any discussion of cortical function must, of necessity, touch upon the question of intrinsic cortical information flow and cortical cell types. However, the available knowledge for the cerebral cortex is insufficient to allow firm circuits to be drawn. Even were this not the case, it is unclear that such diagrams would be particularly helpful. For example, the synaptic organization of the cerebellar cortex has been understood for some time, at a level far in advance of anything one might expect for the cerebral cortex in the foreseeable future. There are the Purkinje projection cells, plus Golgi, basket, stellate, and granule interneurons. The synaptic interrelations of these types are also well known and can be found in many physiology textbooks. However, all of this detailed knowledge of the intrinsic cell types and the inhibitory and excitatory postsynaptic potentials they generate did not suffice to prevent Brooks and Thach (1981, pp. 937–938) from concluding that: "the modes of execution of these [cerebellar] functions are still largely unknown. . . . There are difficulties with all the theories of cerebellar function." So, if detailed diagrams of cerebellar circuitry do not solve the problem of cerebellar functions, the more imperfect knowledge of neocortical circuits may be even less useful in consideration of cerebral cortex physiology. What we hope to accomplish here is to show that what little is known about intrinsic cortical organization is not inconsistent with the sort of neural switching process we discuss elsewhere in this monograph.

If one attempts to classify cells of the cerebral cortex, it is immediately apparent that the number of output cell types discussed in the previous section is rather large. Table 5.1 shows that there are several types of output cells in different laminae in cerebral cortex. Projections to other cortical areas arise in large part from layer III, projections to thalamic main relay nuclei arise mostly from layer VI, and outputs to a whole series of subcortical structures (e.g., the spinal cord, pons, red nucleus, thalamus, and striatum) arise from layer V. Different groups of layer V output elements vary in size and position, and it seem appropriate to assume that there are several different output cell types in the layer. Considering all these, it is apparent that there are at least seven different output cell types in cerebral cortex, even if one lumps together (as a single cell type) all corticocortical cells regardless of where they project within the hemisphere or, via the corpus callosum, to the opposite hemisphere.

In addition to these various projection neurons, all of which are classified as pyramidal, there are a number of morphologically distinct intrinsic local circuit neurons. These are summarized in Table 5.2 and include at least one spiny type, thought to be excitatory, and several smooth multipolar types, most of which are probably GABAergic and inhibitory. At least some cells of one other type, the bipolar cell, is peptidergic (Peters and Connor, 1983), and bitufted cells may be peptidergic, as well.

We do not intend here to review the types of nonpyramidal cells that have been observed in the cerebral cortex. Treatments of this difficult problem in classification include those by Lund (1973), Jones (1975, 1981a), Szentágothai (1979), Valverde (1976), and, most recently, Peters and Regidor (1981). It is generally agreed that these local-circuit neurons can be grouped into some 8–12 classes based on their spine density, dendritic morphology, and pattern of axonal ramifications. Table 5.2 lists the names of these local-circuit neurons, following the dendrite-based classification scheme of Peters and Regidor (1981) for cat striate cortex. Of course, these may vary depending on the cortical field, species, and investigator, but there is general agreement on the overall scheme. For our purposes, the scheme of Peters and Regidor should suffice to enable us to discuss the circuitry that makes plausible short-term input-output switching in the cerebral cortex.

In Chapters 8, 9, and 10 we will examine methods to explore changes in the efficacy of specific input pathways to the precentral motor cortex. If switching of the type we discuss is to occur in the cortex, it must be necessary to suppress certain inputs specifically. One way this could be accomplished for corticocortical pathways is if the output of one cortical

Table 5.2. Nonpyramidal Cells in Cerebral Cortex

Name Based on Dendritic Morphology	Proposed Simple Name
Neurons in Layers II–V	
Smooth, spherical multipolar neurons	Small multipolar cell
Sparsely spinous, spherical multipolar neurons	Sparsely spinous stellate cells
Spinous, spherical multipolar neurons	Spinous stellate cells
Sparsely spinous or smooth multipolar neurons with elongate dendritic trees	Basket cells
Smooth, bitufted neurons with vertically oriented dendritic trees and chandelier axons	Chandelier cells
Sparsely spinous, bitufted neurons with vertically oriented dendritic trees	Sparsely spinous bitufted cells
Sparsely spinous or smooth bipolar neurons with vertically oriented dendritic trees	Bipolar cells
Neurons of Layer I	
Smooth, bitufted neurons with horizontal dendritic trees	Horizontal cells of layer I
Neurons of Layer VI	
Sparsely spinous, multipolar neurons with elongate dendritic trees	Basket cells of layer VI
Sparsely spinous, multipolar neurons with spherical dendritic trees	Sparsely spinous cells of layer VI
Sparsely spinous or smooth bipolar neurons with horizontal dendritic trees	Horizontal cells of layer VI

Source: After Peters and Regidor, 1981.

field to another could be switched off. Well situated to shut off the output of one cortical field to another is the chandelier cell (Figure 5.3D), a type of multipolar neuron that is thought to be GABAergic (Peters et al., 1982), inhibitory, and to form profuse synapses on the initial segment of pyramidal cell axons, especially in supragranular layers of cortex. It could be expected that exciting this cell, via another corticocortical or other extrinsic connection, would cause a chloride shunt so powerful that it would block pyramidal cell output regardless of the inputs to that neuron. By concerted action of a number of chandelier cells, the corticocortical output of an area or part

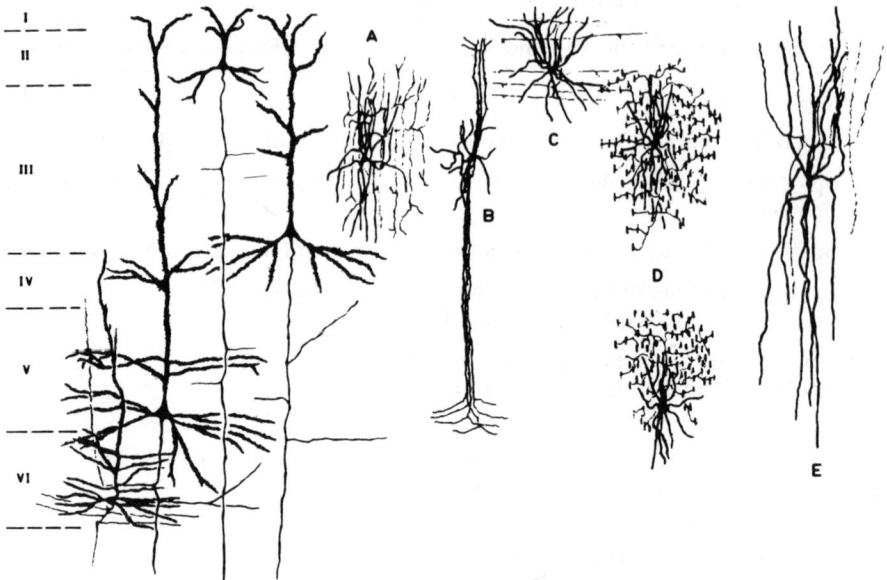

Figure 5.3. *Some of the cell types in the cerebral cortex as revealed by the Golgi methods.* The four cells on the left are: (A) Pyramidal cells. (B) Double bouquet cell. (C) Basket cell. (D) Two chandelier cells. (E) Large bitufted cell. (From Jones, 1981a, figure 16)

of one could presumably be completely turned off (cf. Fig. 8.5). The laminar organization of the chandelier cells, though not completely determined, suggests that chandelier cells are concentrated in superficial layers (Fairén and Valverde, 1980). Although this restriction is not as complete as at first believed (Peter and Regidor, 1981; Jones, 1981a), it seems likely that the effects of chandelier cell (initial segment) inhibition are directed mainly toward corticocortical cells.

A similar role could be performed by the large basket cells (Figure 5.4), another type of GABAergic cell (Hendry and Jones, 1981), one that synapses on pyramidal cell somata. In contrast to the chandelier cells, basket cells distribute their terminals on pyramidal cell bodies and are not restricted to superficial layers. Therefore, it might be imagined that basket cell (somatic) inhibition would be directed toward both corticocortical and corticofugal cells. Taking the basket and chandelier cells together, it is possible to imagine parallel gating circuits, one switching off the output of a cortical field generally, the other eliminating corticocortical (including commissural) outputs specifically. For now it must be emphasized that these formulations are only guess-work, but they should serve to show how readily the output of a cortical column might be turned off, in part or completely.

For a consideration of how input-output switching could be accomplished by the known or suspected cortical circuitry, the route of information into the cortex must be considered. Unfortunately, the technical limitations of acute neurophysiological and neuroanatomical techniques have led most investigators to concentrate on the information-processing events that take place within the first three synapses (or approximately 2 msec) of the signal reaching the cortex (see Ferster and Lindström, 1983; Jones, 1981a; Eccles, 1981); the events of interest to us occur over a period of time at least an order of magnitude larger. Nonetheless, it is generally thought that input to cortex engages the output elements both directly (White, 1978, 1979, 1981) and indirectly via certain of the spiny nonpyramidal cells (Jones, 1975; 1981a). These interneurons are termed spinous stellate by Peters and Regidor (1981) and Lund (1973), and occasionally some of them are termed star pyramids (Lorente de Nó, 1949). It is argued that these spiny nonpyramidal cells relay information from the region that receives specific thalamic information to the pyramidal cells (Lorente de Nó, 1949; Szentágothai, 1979; Jones, 1975, 1981a).

In Chapters 8 and 9 we discuss the possibility that thalamic afferents from two sources—the interpositus and the dentate nuclei—project to at least partially different sets of spiny nonpyramidal cells, and that these pathways can be selectively suppressed or facilitated. Note that if these

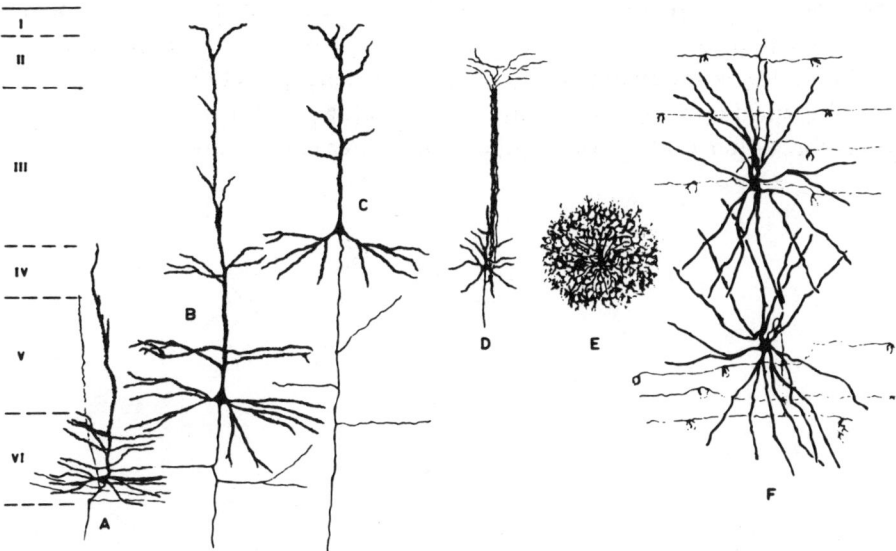

Figure 5.4. *Additional cell types in cerebral cortex.* (A) Probably corticothalamic pyramidal cell. (B) Probably corticospinal pyramidal cell. (C) Probably corticocortical pyramidal cell. (D) Spiny stellate cell. (E) Golgi-type II cell. (F) Two basket cells. (From Jones, 1981a, figure 3)

independent inputs to cortex, one via the dentalothalamic route, the other via the interpositothalamic route, were to terminate on separate populations of the spiny nonpyramidal cells, the efficacy of each pathway's inputs to a population of pyramidal output cells could be regulated independently. The axonal ramifications of the bitufted cell (Figure 5.3B, one of the types of cellule à double bouquet of Cajal), thought to terminate both on the spiny nonpyramidal cells and on pyramidal cell apical dendrites (Jones, 1975), seem well situated to perform precisely this role, that is, regulating the excitability of the spiny nonpyramidal cells and perhaps other pyramidal cell inputs, as well.

In contrast to the cerebellar cortex, in which local circuit neurons have been intensively investigated in acute electrophysiological experiments and a beginning has been made on investigating their roles in sensorimotor processes of behaving animals, virtually no studies of identified local-circuit neurons have been conducted in any cortical area of behaving animals. Only the output elements of the somatic sensorimotor cortex have been identified in such animals. However, we may not need to be distressed by this lack of data about information processing within the cortex. Although the lack of basic electrophysiological data (from acute experiments) on identified local-circuit neurons in cerebral cortex makes it premature to attempt to formulate experiments that would involve studying the relation of local-circuit neurons to behavior, the techniques are available for studying the input-output relations of cortical fields and columns. It should be appreciated that, insofar as the rest of the brain is concerned, the function of the cortex is mediated by its projection neurons. Only by influencing its subcortical targets, such as the spinal motor pools, can the cortex affect behavior, either directly or indirectly via other cortical areas.

Chapter 6

Set and the Single Unit

WIRES IN THE BLACK BOX

We saw in Chapter 2 that the idea of set-dependence in behavior is an old one, and that although there was a period when it was unfashionable to consider the dynamic events in the black box of the brain, many psychophysicists continued to investigate and to speculate about these indirectly observable brain functions. In particular, it was established that *perceptual set* determines, to a substantial extent, the signals that are detected and the processing of the sensory information. Further, it was established that preparation to respond to a stimulus, termed here *motor set*, greatly influences the behavior that occurs in response to a stimulus, as can be observed by a careful examination of the activity patterns of muscles. In this chapter, we discuss some observations, derived from single-unit neurophysiology, that may reflect or serve as the foundation of preparatory set. The link between the single-unit approach to the black box and behavioral observations is necessarily insecure, but it seems to us that, at least in some respects, it may be valuable to treat these two bodies of information within a single conceptual framework.

In our discussions of motor set within this monograph, we have relied on the following experimental paradigm: an instruction stimulus (IS) establishes a motor set which, while not calling for an immediate overt reponse, specifies the response or aspects of the response to a subsequent trigger stimulus (TS). There might be many levels and sources of instruction stimuli. For example, a human subject could be instructed to pronounce, after the ocurrence of a tone (the TS), the sum of three and five. The subject could then establish a motor set for the verbal expression of the sum: "eight." Alternatively, the subject could be instructed to "add the following pair of numbers." Then, after a delay, the subject is given the values to be added. This might be considered a "set for addition," an instruction which operates at a clearly more general level than the set to say the word "eight." Nonetheless, the general paradigm is unaltered; the situations differ only in the extent to which an IS informs the subject of the response to be made or the operations to be performed on subsequent incoming information.

Since the era of Clever Hans, few investigators have attempted arithmetic tasks in nonhuman animals. However, analogous set-dependent tasks have been designed for performance by nonhuman primates and, in association with modern neurophysiological methods, it has been possible to investigate the activity of neurons in monkeys while they performed a variety of these set-dependent behaviors.

The primary difficulty with the literature, besides the mnemonically demonic nature of many of the behavioral tasks, is the unfortunate problem of "area-specific experimentation." This is the phenomenon in which one investigator examines one cortical region with a task designed specifically for that region, while another investigator, in another part of the world, investigates another area during the execution of an entirely different, but conceptually related, task. It is therefore necessary to exercise extreme caution in comparing the results obtained in different laboratories with different tasks in different parts of the brain. We will, however, ignore this sound advice and attempt to summarize some of what has been learned about the activity of neurons in the cerebral cortex during periods of motor or perceptual set. This chapter is not meant to be an exhaustive review of the literature, but rather is intended to support the proposition that there is profound neuronal activity during periods in which a monkey waits to execute a response, and that this activity can be found extensively in the cerebral cortex, if not throughout the entire cortex.

The neurophysiological experiments in this category are obligatorily performed in awake and totally unanesthetized animals, nearly always a species of macaque monkeys. They fall roughly into two overlapping categories: many experiments seek to show the effect of set, usually perceptual set (i.e., selective attention), on the neuronal responses to external environmental stimuli; other experiments are designed, usually in addition to the first objective, to examine the sustained activity after instructions have been presented.

Because so many of these experiments involve instructions, it is important to discuss the variety of different sorts of instructions that might be encountered. As mentioned above, instruction stimuli can be at different levels of generality; additionally, instructions need not derive from stimuli at all. Nonstimulus instructions might include the subject's prior response (as in delayed alternation) or an estimate of time. Concerning the instructions that emanate from external stimuli, they can, of course, belong to any conceivable sensory modality, submodality, or combination thereof. Within this multitude of ISs it is important to distinguish between two major classes: spatial-instruction stimuli and nonspatial instructions. A spatial instruction is one in which the stimulus itself is the target of the instructed response. We term this a *direct* sensorially guided response. These cues most often involve visuospatial stimuli, although auditory or tactile signals can serve the same role. By contrast, instruction stimuli may have no spatial relationship to a target of the motor response. These instruction stimuli we term *arbitrary* and the responses to them arbitrary sensorially guided movements. A simple example is when a red light instructs a movement to the right and a green light instructs a movement to the left.

There have been a number of investigations of neuronal activity during periods when a monkey has been given a stimulus but is withholding an overt motor response. Investigated cortical regions include: prefrontal, inferotemporal, posterior parietal, auditory, supplementary motor, precentral motor, and premotor cortex. In each region, neuronal activity during this "waiting" period is interpreted in the context of the prevailing sensory input or presumed functional specialization of that field. Thus, activity during this waiting period is thought to be associated with visual fixation (Mountcastle et al., 1975) or directed visual attention (Robinson et al., 1978; Motter and Mountcastle, 1981) in the posterior parietal cortex; short-term memory of visual information in the inferotemporal cortex (Fuster and Jervey, 1982; Mikami and Kubota, 1980); spatial mnemonic functions in the prefrontal cortex (Niki, 1974a; Kubota et al., 1974, 1980; Kojima and Goldman-Rakic, 1982; Fuster, 1973); and sensorimotor integration in the auditory cortex (Vaadia et al., 1982). In the premotor (Weinrich and Wise, 1982; Godschalk et al., 1981), supplementary motor (Tanji et al., 1980), and precentral motor (Tanji and Evarts, 1976) cortex, this activity has been linked to motor preparation.

In addition, a number of investigations, including some of those listed above, have examined the effects of instructions on neuronal responsiveness to sensory stimulation. These studies involve the prefrontal cortex (Mikami et al., 1982); frontal eye fields (Bushnell et al., 1981; Bruce and Goldberg, 1981); parietal cortex (Robinson et al., 1978; Goldberg and Bushnell, 1981); inferotemporal (Wurtz et al., 1983) and other parts of extrastriate visual cortex (Fischer and Boch, 1981a, b); auditory cortex (Hocherman et al., 1981; Benson and Hienz, 1978) somatosensory cortex (Hyvärinen et al., 1980); and precentral motor cortex (Tanji and Evarts, 1976; Poranen and Hyvärinen, 1982).

PRIMITIVE SET-RELATED NEURONS

Before embarking on a summary of the experiments performed on monkeys, it might be useful to consider invertebrates and the set-related mechanisms involved in their limited behavioral repertoire. During feeding, withdrawal from tactile stimuli is suppressed in Pleurobranchaea when the sensory stimuli for feeding and withdrawal are presented simultaneously (Kovac and Davis, 1977). This "dominance" of feeding behavior over withdrawal behavior occurs because the central nervous network that controls feeding inhibits the central nervous network that controls withdrawal. Kovac and Davis showed that the inhibition is largely mediated by a bilaterally symmetrical pair of identifiable feeding neurons that are members of a "corollary discharge" population in the buccal ganglion.

In this case, we may think of the IS as food (detected by chemoreceptors) and of TS as a touch to the anterior oral veil. When the animal is set to feed, TS fails to elicit withdrawal, whereas the TS elicits vigorous withdrawal both before and after feeding. In order to establish this causal relation between set-related activity elicited by IS and altered response to TS, Kovac and Davis (1977) systematically stimulated identified neurons in the feeding motor network, one at a time, with an intracellular microelectrode. They discovered a pair of nerve cells in the buccal ganglion with axons that ascend to the brain and powerfully inhibit the withdrawal output. These two neurons normally discharge during feeding and are suppressed during withdrawal. When they are hyperpolarized by an intracellular electrode, thus preventing action potentials, feeding no longer suppresses withdrawal. Conversely, depolarization of these cells to the point at which they discharge at physiological frequencies almost completely suppresses withdrawal under any circumstances.

In the mammalian nervous systems, no individual cell is likely to be critical, and the type of demonstration carried out by Kovac and Davis probably cannot to be duplicated in mammals. In fact, the situation in Pleurobranchaea is now felt to be more complex than originally thought (Kovac et al., 1983a,b; Davis et al., 1983). Their original report can, however, provide a conceptual model for the study of set-related activity. To quote Davis (1979, p. 7) on the way in which his experimental paradigms may serve as a basis for analysis of behavioral hierarchies in general:

> . . . the organization of behavioural acts into a priority sequence, or hierarchy, furnishes a possible conceptual framework for understanding the modification of behaviour by experience, including learning. As noted above, for example, feeding normally dominates withdrawal in the behavioural hierarchy of *Pleurobranchaea*. When specimens are avoidance-conditioned by pairing food with response-contingent shock, however, they rapidly learn to withdraw from the formerly palatable food substance. That is, the animals experience a reversal of the usual hierarchical relationship between feeding and withdrawal. There is some evidence that feeding and withdrawal are mutually inhibitory. Perhaps the negative reinforcement of avoidance-conditioning alters the relative strength

of the respective inhibitory pathways, resulting in the reversal of priorities that is represented by the learned behaviour.

To put this discussion in the framework we have been using in this monograph, we might say that the animal can learn to activate neurons that cause a set for feeding subsequent to the touch (trigger) stimulus. Further, it appears that the animal can learn to inhibit these set-causing neurons, thus causing a set for the opposite (withdrawal) movement after the same stimulus. We can imagine that, in the cerebral cortex, assemblies of neurons may play an analogous role in the preparation for a motor act, and that neural activity during a waiting period plays a variety of roles, depending on the cortical field in which it occurs. These roles may involve some sort of short-term storage of information, for example motor programming, information about environmental events or objects, and selective direction of attention toward aspects of these stimuli, or they may regulate the efficacy of other pathways.

THE FRONTAL GRANULAR CORTEX

There have been many studies of neuronal activity in the frontal cortex of primates. We shall consider two regions separately: the frontal granular cortex, which includes what is commonly termed either the prefrontal, prearcuate, or periprincipal cortex, and the frontal agranular cortex, which includes the precentral motor cortex (MI), the supplementary motor cortex (MII), and the premotor cortex. Other parts of the frontal cortex have been less intensively studied (Thorpe et al., 1983; Niki and Watanabe, 1976b) and are not dealt with here.

Most studies of the frontal granular, or prefrontal, cortex have concentrated on the activity during two conditions: when the monkey has been given an instruction, by sensory cue, of what response must be executed in order to receive a reward, or when the animal must alternate its movements. The former is termed a delayed-response task and the latter a delayed-alternation task, but we will not use this system of nomenclature. Rather, we will concentrate on when and how information that will guide a movement is delivered. In this framework, it can be seen that the difference between the two general types of tasks outlined above is that, in the former, the animal uses cues from the external environment to guide its behavioral responses and, in the latter, uses information about its own past or current behavior. In spite of a set on the part of numerous investigators to search for and study neural activity that seems to be "serving" a "spatial mnemonic function," the published studies indicate that under virtually all experimental conditions a rather large proportion of neurons in the prefrontal cortex show activity modulation well correlated with the responses made by the animal. This fact does not mean that *all* activity in the prefrontal cortex can be accounted for by a relation to movement, however, and many experiments have demonstrated patterns of activity that cannot be easily accounted for on the basis of the motor act executed by the animal.

Two sorts of studies have shown that neurons in the prefrontal cortex are tonically activated before an impending movement of a certain type. This activity follows a stimulus that instructs the animal about its next movement and continues until about the time the movement is begun. In many experiments, two instruction stimuli are given sequentially with delays between them. In these, the first instruction stimulus must be remembered and combined with the second for the monkey to know the precise movement that will be rewarded. Experiments of this type include those by Kubota et al. (1980), Watanabe (1981), Fuster et al. (1982), and Rosenkilde et al. (1981). In other experiments, a single instruction is given that informs the animal about the next rewarded movement. Reports using this type of behavioral paradigm in the frontal granular cortex include those by Fuster (1973); Kubota and Niki (1971); Kubota et al. (1974); Niki (1974c); Niki and Watanabe (1976a, 1979); Kojima et al. (1981); Komatsu (1982); Kojima and Goldman-Rakic (1982); and Vaadia, Benson, and Goldstein (unpublished data). Finally, a third type of procedure involves alternation between two targets (Niki, 1974a, b). Certain studies have combined two of these approaches while studying single neurons.

Three basic patterns of neuronal modulation have been observed in neurons while monkeys performed these sorts of tasks: transient modulations correlated temporally with the (usually visual) instruction stimulus; transient modulation correlated with the execution of movement; and sustained changes in activity, usually following an instruction stimulus and lasting until about the time of the overt motor response. A variety of speculation has been based on these observations, often regarding a role for these neurons in visually guided movements or spatial memory. We do not deal with those speculations here; rather, we concentrate on some of the properties of the transient and sustained activity changes that follow an instruction.

A sustained modulation of neuronal activity after an instruction was first observed by Fuster and Alexander (1971) and by Kubota and Niki (1971). Although Fuster (1973) did not find any cells that showed activity changes specific for those trials in which a particular motor response was to be made, Niki (1974a, c) did find such *directionally specific* neurons in the frontal granular cortex. Thus, many of the neurons showing sustained activity change after an instruction show this change only before movements in one direction. Other single units demonstrate directional specificity by showing increases in activity before movements in one direction and decreases before movements in the opposite direction. This finding alone suggests that neuronal activity might underlie the preparation for specific motor responses or, alternatively, might reflect the instructions that elicit those responses. These we call "set-related" neurons. It is interesting to note that sustained modulations are similar when a visuospatial stimulus instructs the animal about its next movement (Kubota and Niki, 1971; Fuster, 1973); when the monkey's preceding movement serves as the instruction for the next movement (Niki, 1974a); when the monkey awaits a subsequent trigger stimulus (Niki and Watanabe, 1976a); or when the monkey times the delay without an external trigger stimulus (Niki and Watanabe, 1979; Kojima et al., 1981). In general, it has been reported that the prefrontal cortex contains neurons that appear to be specific for both the instruction stimulus itself (or aspects of the stimulus or instruction) and for the motor set.

Watanabe (1981) trained monkeys to respond with a movement to a rightward or a leftward target depending on two visual cues, the second delivered one second after the first. The first visual stimulus had to be remembered for the one second and combined with the second cue (the IS) for the monkey to know the target to be rewarded. After *another* delay period, the monkey was given a trigger stimulus and allowed to execute together the movement indicated by the two visual cues. Though this task might seem difficult, Watanabe (1981) reported that the monkey, once well trained, performed the task almost perfectly. It is possible, then, to separate sustained activity that might be considered a "response" to either the first (visual) or second (visuospatial) instruction stimulus, and that related to the impending motor response. Many of the units showed little or no specificity for different instruction stimuli or different motor acts, but instead showed rather nonspecific modulation during the task. However, of those neurons that showed specific patterns of activity, the vast majority (73%) were related, according to Watanabe's analysis, to the direction of the motor response to be executed. Most (such as that illustrated on the left of Figure 6.1) showed sustained activity after the second instruction stimulus, the one that gave the monkey the final chunk of information necessary to execute the proper movement; some showed transient activity shortly preceding the onset of movement. Much of the sustained neuronal modulation in the prefrontal cortex therefore would appear to be specific either for preparation for movement, or for the specific instructional significance of the stimulus. These two possibilities are not easily separated, and few experiments have been designed to do so.

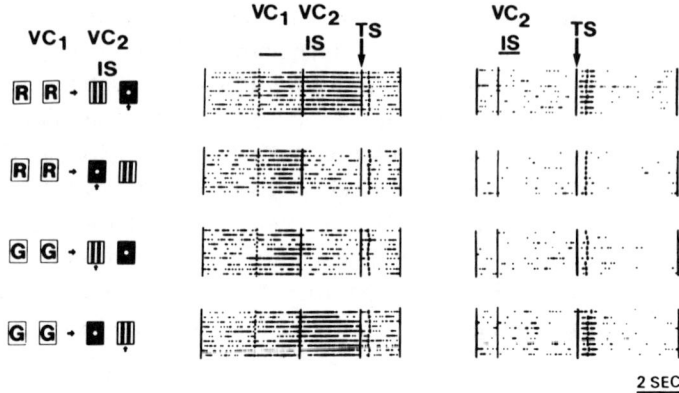

Figure 6.1. *Effects of instructions on cell discharge.* In this experiment on prefrontal cortex, the monkey is given two sets of visual stimuli or cues. Left column: The first visual cue (VC_1) is either red (R) or green (G). The second visual cue (VC_2), which serves as the instruction stimulus (IS), is presented later and consists of two targets, one indicated by the circle, the other by the vertical stripes. A red VC_1 means that the monkey should respond by hitting the target indicated by the circle, and a green VC_1 means to hit the striped target. The correct response is indicated by the upward-pointing arrow in each row. Raster displays of two prefrontal cortex units are shown in the middle and right columns. The middle column shows sustained activity following the instruction stimulus (IS) that is specific for the direction of the response (i.e., the target being to the right), rather than the IS guiding that response. The right column shows a transiently activated cell also specific for the direction of movement. (From Watanabe, 1981, figure 5)

Similar conclusions derive from other studies of the frontal granular cortex (e.g., Niki and Watanabe, 1976a; Kubota et al., 1980). Niki and Watanabe compared sustained activity that followed two types of instruction stimuli. In the first type, the monkey was instructed by a visuospatial cue (for example, the leftmost of two targets) to make an analogous movement (in this example, a leftward movement to the target). Similarly, the rightmost instruction stimulus meant for the animal to move his limb to the right, the uppermost of two stimuli meant he was to move up, and the lower of two stimuli meant he was to move down. This situation of direct (see above) instruction-response relationships was contrasted with a situation in which the relationship was arbitrary. For example, the upper of two instruction cues signified that the rewarded movement would be to the right and the lower of the two instructions meant that the animal was to move left. In this way the "specificity" of sustained unit activity with the instruction stimuli or the impending motor responses could be evaluated independently.

Two groups of directionally specific neurons were found. In the first group, a sustained modulation was observed following a given instruction stimulus and the neuronal activity was similar, regardless of the movement it instructed the animal to make. The sustained activity of neurons in this group was approximately the same when the lower of two instruction stimuli served as an instruction for downward movement as when it instructed a leftward movement. This pattern of activity, seen in about three-fourths of the appropriate and tested units, seems to be a reflection of the instruction stimulus itself. However, another class of neuron was observed, with activity appearing to reflect the motor preparations of the animal. In these neurons, approximately the same pattern of activity was seen when either of two possible instruction stimuli cued a given movement — that is, when the *rightmost* of two stimuli instructed a rightward movement or the *uppermost* of two stimuli instructed the same rightward movement, the neurons showed a similar pattern of modulation, usually a sustained increase in discharge frequency. Figure 6.2 shows one of these single units. The conclusion that activity of these cells was specifically related to motor set was reinforced by the finding that before an erroneous movement the cell did not show the characteristic set-related activity. Of course, such a conclusion must not be accepted uncritically, since a number of important variables (e.g., eye movements) have not been controlled. In most studies such as those cited above, it has been assumed that the sustained activity is not simply to eye movements or visual fixation. An exception was reported by Suzuki and Azuma (1977), who found that some prefrontal cortex units were correlated with visual fixation of a trigger stimulus. Less commonly has the question of postural adjustment been addressed. It remains possible that some of the sustained activity reflects postural (or other motor) activity on the part of the animal while it awaits a trigger stimulus. Internal controls in the cited experiments argue against this possibility, however. For example, in the experiments of Niki and Watanabe, cited above, sustained activity changes observed in some cells depended on the visuospatial stimulus, independent of the motor response that the stimulus instructed, but postural adjustments should depend upon the motor response.

It is especially in this context that the report of Niki (1974b) is of interest. The experimental situation is shown in Figure 6.3. He found that for neurons with directionally specific sustained activity, the activity was in large part independent of the actual movement the animal was to make. Rather, the activity appeared to signal the relative location in space of two potential target stimuli. Four targets were placed in front of the monkey and any two of these might be illuminated on a given block of trials. The monkey's task was to alternate between the leftmost and the rightmost of the two illuminated targets. In a sense, it can be said that the instruction derived from the animal's previous responses. By changing the targets, it was

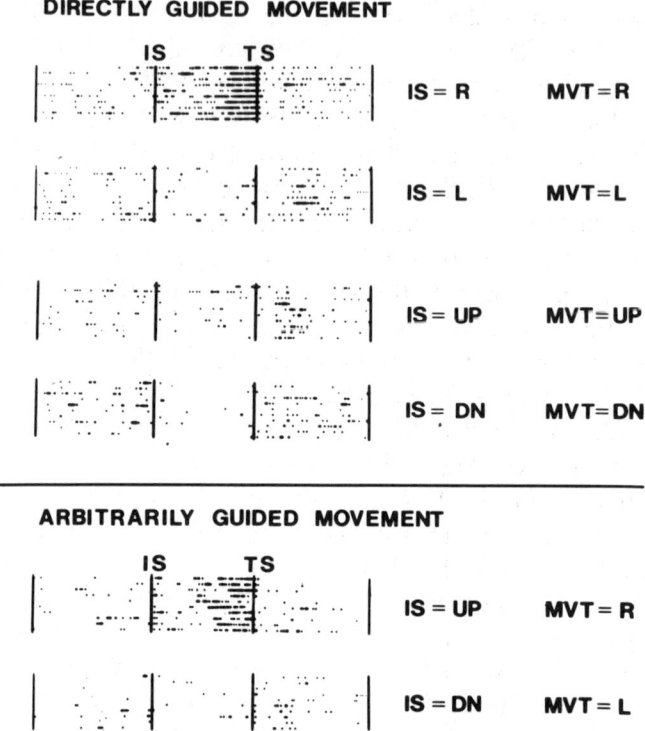

Figure 6.2. *Motor preparatory selectivity is seen in this prefrontal cortex cell,* the activity of which relates to the direction of response rather than the visuospatial signals (IS) that guide the response. The monkey was trained in two tasks. The upper four rasters show single-unit activity during *directly* guided movements (MVT) to the right (R), left (L), up (UP), and down (DN). The bottom pair of rasters shows activity during an *arbitrarily* guided movement. In this task, the upper visuospatial cue (IS) instructs a movement (MVT) to the right target and the lower cue a movement to the left. Note that the unit activity following the instruction stimulus (IS) is only seen before responses to the right, regardless of the instruction cue. (From Niki and Watanabe, 1976a, figure 4)

possible to make a given target either leftmost or rightmost. All of the testable neurons that showed sustained activity changes reflected the relative position of the targets.

If Niki (1974b) was observing the same type of cell as in the other experiments described above (Niki and Watanabe, 1976b), then these findings argue against the contention that activity in such cells reflects a postural adjustment on the part of the animal preparing to make a movement to a specific target. Similarly, findings in the frontal eye fields, another part of the frontal granular cortex, argue against postural adjustment mediating all sustained preparatory activity, because posture is not a factor in preparation

THE FRONTAL GRANULAR CORTEX

for eye movement. In the frontal eye field, it is presumably easier to rule out postural factors affecting single-unit activity. Bruce and Goldberg (1981) reported that, in the frontal eye field, sustained activity preceded saccadic eye movements "if the task repeatedly required a certain saccade. This anticipatory discharge ceased when several trials not requiring the saccade were given." If these neurons are specifically related to eye movement, postural adjustment would appear to be ruled out. The frontal eye field was also found to contain neurons discharging for several seconds after certain saccades until the next one, as if holding a short-term memory of the most recent eye movement (Goldberg and Bruce, 1981).

Transient activity changes after instruction stimuli follow similar patterns of activity. Most prefrontal units appear to be modulated specifically during situations in which the response is guided by an instruction stimulus and

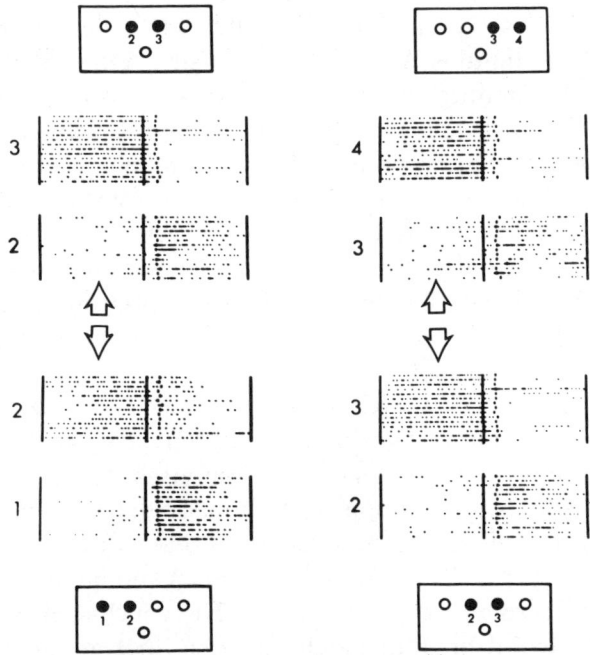

Figure 6.3. *Prefrontal discharge coding relative target position.* Near each pair of raster displays is a diagram of the visual targets seen by a monkey whose task was to alternate between the leftmost and rightmost of two illuminated keys (indicated by the filled circles). The rasters show the neuronal activity associated with responses to the numbered keys. The salient comparison is between responses to key number 2 when it is leftmost (upper left) and rightmost (lower left) or between responses to key number 3 when it is leftmost (upper right) and rightmost (lower right). Note the profound difference in activity depending on the relative location of the illuminated target, rather than its actual position in space. (From Niki, 1974b, figures 1 and 2)

not when the same stimulus is given under conditions when it is behaviorally irrelevant to the animal. Vaadia, Benson, Hienz, and Goldstein (unpublished data; see also Sakai, 1974) examined neuronal activity in each of four conditions: when the monkeys were guided by direct *visuospatial* instruction stimuli to make movements to one of five targets; when direct *auditory* cues served the same instructional purpose; when the identical auditory cues served only as a trigger stimulus, the instruction being provided by the prior response; and when the animal was not performing the task. Although there were exceptions, most of the prefrontal neurons that were transiently activated after an auditory signal that instructed for a particular movement were also activated after the visual signal that instructed the same movement. Importantly, when the identical stimulus served only as a trigger (in this case, an advance trigger stimulus since the animal had to withhold response for an additional 500 msec), most neurons showed little or no modulation in "response" to the auditory stimulus. An interesting comparison could thus be made in this experiment. The unit activity could be studied in two situations: when stimulus A served as an instruction to make movement A; and when the identical stimulus A served only as an advanced trigger stimulus and the instruction to make movement A was provided by the animal's immediately prior motor experience. In one case that was studied extensively, the unit activity was specific for the occasions when the stimulus served as an instruction, and did not become activated when it served as a trigger stimulus, though the motor response specified was the same in both cases. It seems possible that at least some prefrontal neurons reflect responses to stimuli that have a specific instructional significance. Other units may be reflecting motor set per se.

As mentioned above, set-relationship studies have taken two forms: examinations of sustained activity that might be involved in the establishment of sets, and the effects of set on neuronal responses. The latter studies have examined the effects of attention, operationally defined, on the transient and sustained modulations. There has been one study of this type in the prefrontal cortex generally (Mikami et al., 1982) and another in the frontal eye fields (Goldberg and Bushnell, 1981). The conclusion in the study of Mikami et al., based on four single units, is that attention to a visual stimulus causes an enhancement of the sustained "response." The study of Goldberg and Bushnell, based on a considerably larger sample, shows that attention alters the transient responses to visual stimulation. This experiment was designed to control the animal's attention by dividing trials into two conditions. In one "attentional" condition, the monkey fixated a spot of light and, after a delay, an extrafoveal visual stimulus served as the target for an immediately produced saccade. Thus the location of the

THE FRONTAL GRANULAR CORTEX 77

extrafoveal stimulus served as an instruction and triggering cue for an eye movement. This was contrasted with a "nonattentional" condition in which the second, nonfixated visual stimulus was irrelevant to the animal's behavior. In this case, dimming the fixation light served as the trigger stimulus for a repetitive limb movement. The result in the frontal eye field was that the visual response to the extrafoveal visual stimulus was potentiated only if it was to be the target of a saccadic eye movement. One of these units is shown in Figure 6.4. Interestingly, attending to an extrafoveal stimulus did not potentiate the response if that stimulus served as the target of an arm movement or if dimming of that stimulus served as the trigger of a repetitive limb movement.

The conclusion that can be drawn provisionally from the single-unit studies on the frontal granular cortex is that there are units which change activity during a period of time when the monkey has received an instruction about its next motor response, but is withholding that response pending the passage of a certain period of time or the arrival of a trigger cue. Speculation about the functional significance of this activity has centered on short-term storage of information about the stimuli that will guide the movement (Fuster, 1973; Fuster et al., 1982), motor set (Niki and Watanabe, 1976b; Watanabe, 1981), visual fixation (Suzuki and Azuma, 1977), and

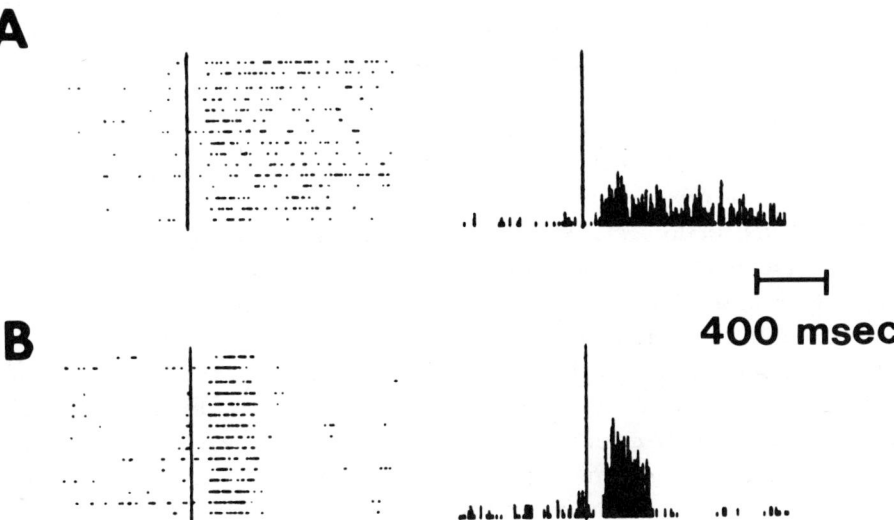

Figure 6.4. *Enhanced response of a neuron in the frontal eye field when the stimulus is the target for a saccadic eye movement.* A: The stimulus, presented at the time of the vertical bar, is irrelevant to the monkey's eye-movement behavior. B: The visual signal is the target of an eye movement. Note the much greater modulation of neuronal activity in B. (From Goldberg and Bushnell, 1981, figure 4)

storage of information about a previously executed movement (Goldberg and Bruce, 1981). Much work of this descriptive sort needs to be done in order to clarify these issues further, but a substantial beginning has been made.

THE FRONTAL AGRANULAR CORTEX

The frontal agranular cortex is immediately caudal to the frontal granular cortex. In it lie cortical fields collectively referred to as the "motor cortex," although the reader should take caution not to interpret this terminology too literally. The frontal agranular cortex differs from the frontal granular cortex in many ways other than cytoarchitecture. For historical, technical, and conceptual reasons, the cortical fields within the frontal agranular cortex have been much better defined than those of the rostrally adjacent granular cortex, although the beginning efforts have now been made in the prefrontal cortex (Aou et al., 1983). At least three main cortical fields have been described: the precentral motor cortex, the supplementary motor cortex, and the premotor cortex (here taken to exclude the supplementary or any other medially situated part of the frontal agranular cortex). There may be additional fields within this general area, most notably a field in the ventral bank of the cingulate sulcus, and some of these fields may eventually be subdivided (Strick and Preston, 1982; Muakassa and Strick, 1979), but for our purposes we will consider only the three main cortical fields.

Sustained effects of instruction stimuli have been described in all three of these motor fields. The first description was, appropriately enough, in the primary, or precentral, motor cortex. Tanji and Evarts (1976) described sustained instruction-related activity in this cortical field. They gave the monkey a visual signal (color) that served as an instruction to push or pull a handle. Many cells, some of them identified as pyramidal tract neurons (PTNs), showed sustained activity changes after these instructions were given, and most were specific for the direction of the upcoming stimlulus. Although not many errors were made by the monkey, Evarts and Tanji reported that when an error was made, the discharge of the unit corresponded to the impending movement rather than to the prior instruction. Figure 6.5 shows one of these units and its activity during a trial in which the monkey performed the wrong response. Kubota and Funahashi (1982) have subsequently observed similar but less striking phenomena, and Kubota and Hamada (1979) less specific preparatory activity in the precentral motor cortex. Kubota and Funahashi noted, however, that in 9 of the 11 neurons that show some sort of an instruction effect, such activity is present only

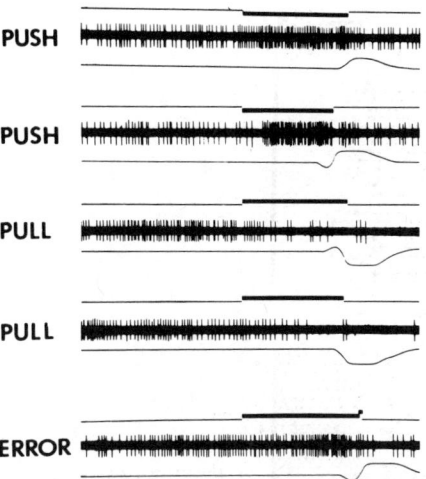

Figure 6.5. *Set-related neuronal activity in the primary motor cortex.* The time of instruction presentation is indicated by the thickening of the horizontal bar above each electrode trace; the instruction delivered is at the left. Below each electrode trace is an indication of the animal's arm position. Push instructions lead to increases in activity and pull instructions lead to decreases. Note that in the bottom record the monkey responds incorrectly (he was instructed to pull, but instead pushes the handle), and in this case the activity resembles other trials in which the monkey ultimately pushes the handle. Thus, the activity appears to reflect the animal's (inferred) motor set, regardless of the instruction. (From Tanji and Evarts, 1976, figure 5)

when the stimuli serve as an instruction and not when similar stimuli serve only as ready signals (the instruction and triggering stimuli in the latter case were delivered simultaneously and later).

In addition to these sustained, set-related unit activity patterns, the effect of set on responses to stimuli could also be examined. Subsequent to the instruction stimulus, the monkey was given a trigger stimulus that consisted of a perturbation of the handle. This was the signal to begin the instructed movement, and also provided a kinesthetic stimulus. The transient neuronal response to this kinesthetic stimlulus was observed to have two temporally distinct components: first, a 20–25 msec latency "reflex" component that was not markedly affected by the instruction stimulus and, second, a 40–70 msec latency "intended" response that was strikingly influenced by the instruction (see Chapter 3). In these neurons, many of them pyramidal tract neurons, the later, intended transient response to the kinesthetic stimulus depended on the direction in which the monkey was instructed to move the handle, and not on the direction of the stimulation. This can be seen in Figure 6.6 for a pyramidal tract neuron in the precentral motor cortex that discharged with pushes of the handle and was silenced during and before pulls of the handle. When the trigger stimulus opposed pushing,

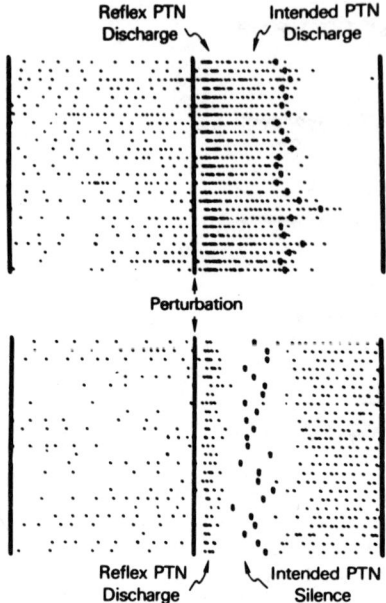

Figure 6.6. *Activity of a pyramidal tract neuron in the primary motor cortex before and after a kinesthetic triggering stimulus (perturbation).* The effect of set on this transient activity is shown. The top rasters show the activity when the monkey was set to push, the bottom when it was set to pull the handle. (From Evarts and Tanji, 1976, figure 3)

the short latency component was somewhat greater than when the trigger stimulus assisted the pushing, but the later component was markedly larger when the monkey had to overcome the kinesthetic stimulus than when he could allow the kinesthetic stimulus to move him in the "proper," or instructed, direction. Evarts and Tanji (1974, p. 479) concluded that these data showed "that motor cortex reflexes can be 'gated' on or off by the voluntary 'set' of the monkey." They also speculated that one of the roles of the sustained, or set-related, activity among precentral and postcentral (somatosensory) projection neurons might be to control or to regulate the kinesthetic input to the spinal cord and cerebral cortex (Tanji and Evarts, 1976, p. 1062): " . . . changes in PTN activity with 'intention' or 'motor set' provide a mechanism for suprasegmental control and presetting of spinal cord reflex excitability specific to the nature of the impending movement."

Recent studies by Poranen and Hyvärinen (1982) also show an important effect of set on precentral motor cortex responses to somatosensory stimuli. In their experiment, clusters of neurons, rather than single units, were recorded. Nonetheless, a profound enhancement in response of the unit clusters could be seen in the circumstance where a vibratory stimulus triggered the movement, as compared to a situation in which it was behaviorally irrelevant. This occurred although the responses triggered by the vibration were executed by the hand opposite to that which received the

THE FRONTAL AGRANULAR CORTEX   81

stimulus, and the response enhancement was seen in the hemisphere that does not control the responding limb.

Set-related activity has also been observed in the supplementary motor cortex by Tanji et al. (1980). About half of these set-related neurons showed a directional specificity, and Tanji et al. (p. 60) reported that the set-effects varied "in relation to enhanced motor skill as the monkeys gained more experience in responding to the trigger stimulus." They also reported a difference in set-related activity when the monkey can predict the direction of the kinesthetic triggering stimulus. Figure 6.7 shows one of these supplementary motor cortex neurons. When the instruction was to push the handle, the cell usually showed a set-related decrease in activity. However, when the monkey "knew" (because of several consecutive identical trials) that the kinesthetic trigger stimulus would assist the movement, the set-related activity was not observed. Thus, the set-related activity in the supplementary motor cortex appears to be specific for the upcoming trigger stimulus, when it, as well as the direction of impending movement, can be predicted. Tanji et al. (1980) interpreted this finding to support a negative-feedback model of kinesthetic function in the motor cortex. They assumed, based on many studies (Conrad et al., 1975; Evarts and Tanji, 1976; Fetz and Cheney, 1980) that if a motor cortex PTN is activated in relation to movement of a particular muscle group, to oppose the action of those muscles by stretching them (via an external kinesthetic stimulus) would excite the pyramidal tract neuron (PTN), presumably causing a muscle action that would oppose or tend to compensate for the kinesthetic stimulus. Tanji et al. speculated that the set-related activity in the supplementary motor cortex might inhibit this negative feedback loop. The activity of the unit shown in Figure 6.7 is consistent with this concept. When the monkey

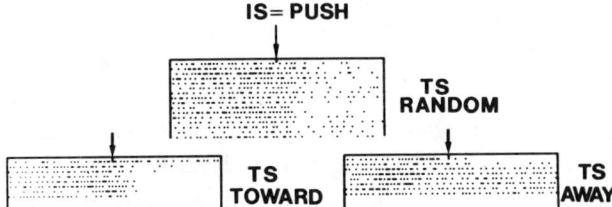

Figure 6.7. *Dependence of set-related modulation on the predictability of a subsequent kinesthetic triggering stimulus.* In all three rasters, the instruction was to push the handle. In the upper raster, the kinesthetic stimulus (TS) was randomized. In the lower two rasters, 16 consecutive trials had the identical kinesthetic stimulus. Note that when the triggering stimulus can be expected to assist the intended movement (pushing *away* the handle) the set-relationship disappears after two or three trials. (From Tanji et al., 1980, figure 3)

either did not know what sort of kinestheic stimulus to expect or expected the trigger stimulus to oppose his intended movement, the unit decreased its activity after the instruction. This, in the view of Tanji et al., might disinhibit the negative-feedback loop described above, allowing the maximum compensation for the opposing handle-jerk. However, when the monkey expected the handle to assist his intended movement, the cell activity remained constant, presumably because disinhibition of the feedback loop was of no value in that circumstance. Translating this suggestion to a population of similarly behaving neurons, one can appreciate that such a predictive function is one of the possible mechanisms that might be performed by set-related activity in the cerebral cortex. Regardless of the validity of how they are interpreted, the phenomena described by Tanji et al. serve as evidence that the set-related activity in the supplementary motor cortex is predictively related to pending trigger stimulus and therefore is not likely to be a simple reflection of the either the visual or the instructive nature of the stimulus.

Transient neuronal activity modulations following trigger stimuli of varying sensory modalities have been studied by Tanji and Kurata (1982) for supplementary motor cortex neurons. They report that many neurons showed activity only when a given trigger stimulus triggered a movement and not when the same (e.g., auditory) trigger stimulus was delivered shortly after the first and did not trigger a movement. These neurons seemed to respond to the auditory stimulus, but only when it served to trigger a movement. This phenomenon is similar to those to be discussed elsewhere in this chapter, especially for auditory cortex, where it has been reported that specific auditory stimuli cause responses only when they trigger one of two possible responses (Vaadia et al., 1982).

Conceptually related studies have been conducted in the premotor cortex, the part of the frontal agranular rostral to the precentral motor cortex and lateral to the supplementary motor cortex. In these studies (Godschalk et al., 1981; Weinrich and Wise, 1982; Wise et al., 1983), similar in many ways to those conducted on the frontal granular cortex, instruction stimuli (usually visuospatial) were given and the monkey was required to withhold movement until a subsequent triggering cue. In the premotor cortex, 34% of the neurons showed a sustained increase in their activity following a stimulus that instructed the animal about what forearm movement has to be executed in order to receive a reward. This activity continued until shortly before the animal began a movement. Such activity was usually selective for the direction of the upcoming movement and terminated in relation to the onset of movement, whether the movement had been cued by a trigger cue. In view of these findings and (a) since this activity was usually similar when an auditory instead of a visual cue provided the instruction, and (b) did not depend on steady visual fixation, on postural adjustment, or on eye movement or eye position, Weinrich and Wise (1982) suggested that in the premotor cortex this neuronal activity reflects the motor preparations or set of the animal.

Further evidence supporting this suggestion has been obtained by Wise et al. (1983). In their experiment, two types of instruction stimuli were delivered to the monkey: one signaled the direction and amplitude of a forearm movement to be executed upon receipt of a trigger stimulus, the other instructed the monkey to withhold a movement. Figure 6.8 shows the activity of a set-related neuron during these two conditions. The premotor

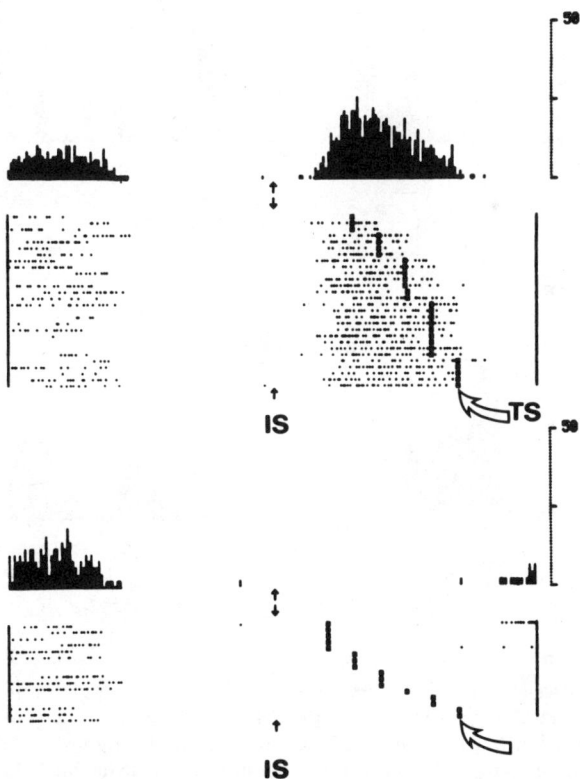

Figure 6.8. *Set-related activity in the premotor cortex.* At the top are trials in which the monkey is instructed (IS) to make an arm movement in a certain direction and later (at the time indicated by the square) is given a triggering stimulus (TS) that allows him to execute the movement. In the bottom raster and histogram, the same cell shows no activity when an identical visual stimulus instructs him to withhold movement. (From Wise et al., 1983, figure 2)

cortex unit became active after an instruction to flex the forearm; no changes were observed after the instruction to withhold movement. This finding supports the idea that at least many of these set-related cells show activity changes specifically related to motor preparations.

A different experiment was performed by Mauritz and Wise (unpublished observations) who trained a monkey to perform a visually guided forelimb movement after direct visuospatial instructions were given. Figure 6.9 shows the set-related activity that occurs when the instruction is altered before the monkey is allowed to execute the movement. The unit activity patterns following instructions to make a leftward movement differ markedly from those seen when the monkey prepares to make the opposite movement.

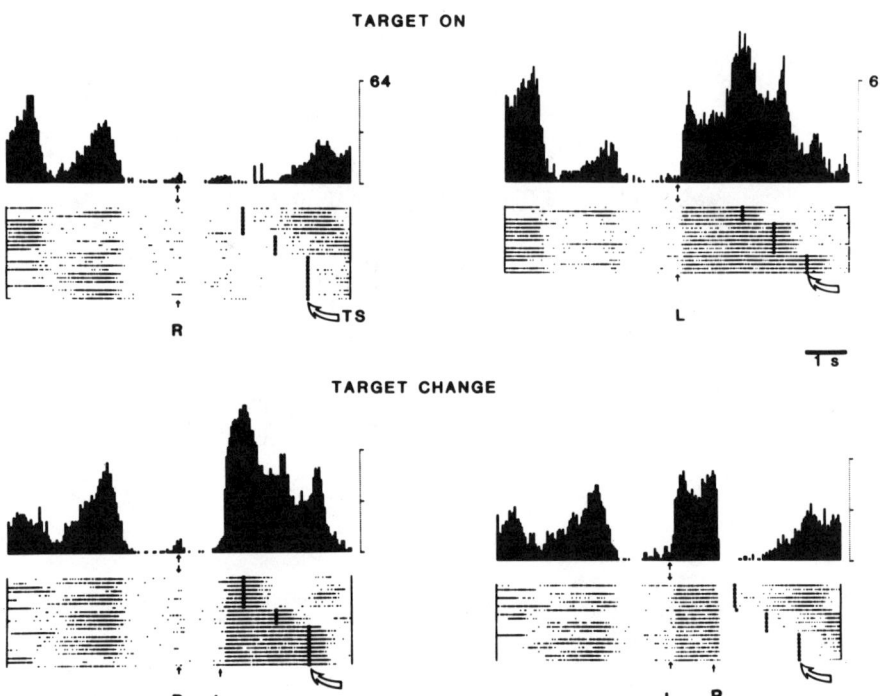

Figure 6.9. *Set-dependent activity in premotor cortex.* This cell rapidly adjusts its set-related activity to reflect changes in the instruction delivered to the monkey. Left column: the monkey is given an instruction (at the arrow) to move his arm to the right (R) and is subsequently given a triggering signal (TS, square). Right column: the instruction is to the left (L). Top row: the cell is excited when the monkey is preparing to move left and inhibited when the monkey is set to move right. Bottom row: the instructions are switched during the period of withholding movement, either from left to right (bottom right) or right to left (bottom left). Note that the cell's activity rapidly changes to reflect this change in instruction. (Mauritz and Wise, unpublished data)

If, however, the instruction is changed from one calling for a leftward to one calling for a rightward movement, the units quickly change their activity from the pattern seen during leftward motor sets to that seen during rightward sets. This change occurs shortly after the second instruction stimulus is delivered. The opposite change in set-related activity is observed when the instruction changes from right to left. This observation, then, also is consistent with the idea that certain of the premotor cortex neurons reflect motor preparation.

This motor set might be envisioned to play a role analogous to that proposed by Tanji et al. (1980) for set-related units in the supplementary motor cortex, and perhaps this influence is mediated via the corticocortical connections that link the premotor and supplementary motor cortex to the precentral motor cortex (Pandya and Vignolo, 1971; Muakassa and Strick, 1979). It also possible that these are corticofugal neurons, and play a role in suppressing subcortically mediated stimulus-response circuits. Whatever the functional significance of these sustained neuronal activity patterns, it appears that they occur ubiquitously in the frontal agranular cortex.

THE PARIETAL SOMATOSENSORY CORTEX

Much less is known about the sustained effects of instruction stimuli in the postcentral somatic sensory cortex. As mentioned above, a small proportion of postcentral pyramidal tract neurons show set-related activity (Tanji and Evarts, 1976). The effects of perceptual set on transient responses to somatic sensory stimuli have been examined by Hyvärinen et al. (1980), who reported that in monkeys the response of single neurons is enhanced when the animal "attends" to the part of the body that is to receive the stimulation.

Mountcastle et al. (1975) did not report any sustained neuronal activity in area 5 of the posterior parietal cortex of the type considered here as set-related. However, they did report a class of neurons whose transient activity was said to be dependent on the set of the animal. These two classes, termed "arm-projection" and "hand-manipulation" neurons, were active in relation to movement of the contralateral limb, but only if the movement was aimed at securing an object the monkey wanted (such as reaching for food or operating a lever that would result in a reward). They observed that these neurons were silent during other sorts of movement (such as those of an aggressive nature) that seemed to involve the same muscles, and that the activity of these neurons could not be accounted for on the basis of sensory input. It was proposed that they performed a "command" function for limb movement under certain motivational conditions, so the activity was dependent on the animal's state — a set-related phenomenon.

THE PARIETAL VISUAL CORTEX

In their initial study of area 7, Mountcastle et al. (1975) reported a class of units activated during the period between the receipt of a visual instruction stimulus and a trigger stimulus. They concluded, on the basis of an examination of the eye movements made by the animal, that these units were specifically related to the act of visual fixation on the stimulus that served as a target, and termed those cells "visual fixation neurons." Robinson et al. (1978) examined the posterior parietal cortex from a different perspective and concluded that cells in that area are not exclusively related to visual fixation, but rather are responsive to the visual stimulus itself. It remains to be determined whether Mountcastle et al. isolated a group of neurons specific for visual fixation and that Robinson et al. missed that group, or whether the later investigators applied adequate stimuli to the cells under conditions not employed by the former group.

There is better agreement on the effects of set on "light-sensitive" neurons. Bushnell et al. (1981) have concluded that when monkeys must attend to an extrafoveal stimulus, the transient response to that stimulus is enhanced. This happens when the extrafoveal stimulus is the target for an arm projection or eye movement (Figure 6.10). Such situations can be compared to one

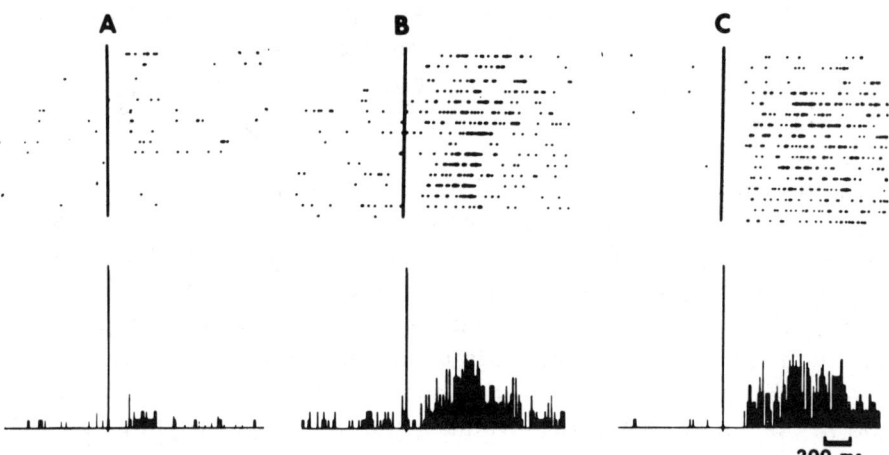

Figure 6.10. *Enhancement of visual responses in the posterior parietal cortex.* In A, the visual stimulus is irrelevant to eye or arm movements. In B, it is the target of a saccadic eye movement, and in C, the same visual stimulus is the target of an arm reaching movement. Note the increased activity in B and C. (From Bushnell et al., 1981, figure 9)

in which the extrafoveal stimulus is behaviorally irrelevant to the animal. This phenomenon contrasts with the enhancement effect in the frontal eye fields (Goldberg and Bushnell, 1981), where the enhancement of the transient response to the stimulus was only observed when the visual stimulus was the target of a saccadic eye movement. Thus, the perceptual set of the animal appears to influence the responses to afferent information. Similar phenomena have been observed in the extrastriate visual fields (Fischer et al., 1981; Fischer and Boch, 1981a, b), but have not been observed in all regions of visual cortex (Wurtz and Mohler, 1976; Wurtz et al., 1983).

TEMPORAL VISUAL CORTEX

The opposite effect on transient visual responses has been observed in the inferotemporal visual cortex. When the monkey was required to respond to a change in brightness of the extrafoveal stimulus, the modulation of inferotemporal neurons decreased, rather than increased, the visual response of the cell (Wurtz et al., 1983). Perceptual set, then, does not uniformly *increase* neuronal responses to sensory stimuli.

Sustained increases in activity following visual (color) instruction stimuli have been observed in the inferotemporal cortex (Mikami and Kubota, 1980; Fuster and Jervey, 1982). In these studies, the authors concluded that the sustained activity is involved in the short-term retention of visual information. Much additional work must be done to substantiate this notion, but it is interesting to speculate that, in certain areas of the cortex, the sustained activity that follows an instruction holds information about that stimulus, whereas in other, presumably motor, areas the activity retains information about the impending motor response.

THE AUDITORY TEMPORAL CORTEX

As in other cortical fields, the activity of neurons in the primary and nonprimary auditory cortex depends in part on the perceptual set of the animal. Benson and Hienz (1978) and Hocherman et al. (1976, 1981) have concluded that attention to or correct prediction of an auditory stimulus has an effect on the auditory cortical response (see also Beaton and Miller, 1975). This effect is not necessarily an enhancement of the response (Hocherman et al., 1981). In the experiment of Hocherman et al., monkeys were trained to respond with one arm to a noise stimulus and with the other arm to a tone stimulus. These served as trigger stimuli as well as instructions. On some trials the noise or tone was the only stimulus present. However, on most trials the auditory stimulus was preceeded by the visuospatial stimulus, which also instructed the animal about its next movement, pending the auditory stimulus. In relatively rare instances, this visuospatial signal *misinformed* the monkey about the pending movement and auditory signal. Thus, neuronal activity could be compared in a situation in which the monkey expected to receive one auditory stimulus or when it expected a different auditory signal. Both of these situations could be compared to the condition in which the monkey received no visuospatial "instruction" at all. That the monkey was making the desired use of the visual stimuli, in spite of the redundant instructional condition, was attested by the monkey's reaction times under the various conditions. When properly instructed, the monkey performed the movement with a much shorter reaction than when misinformed. In this task, two types of changes could be observed:

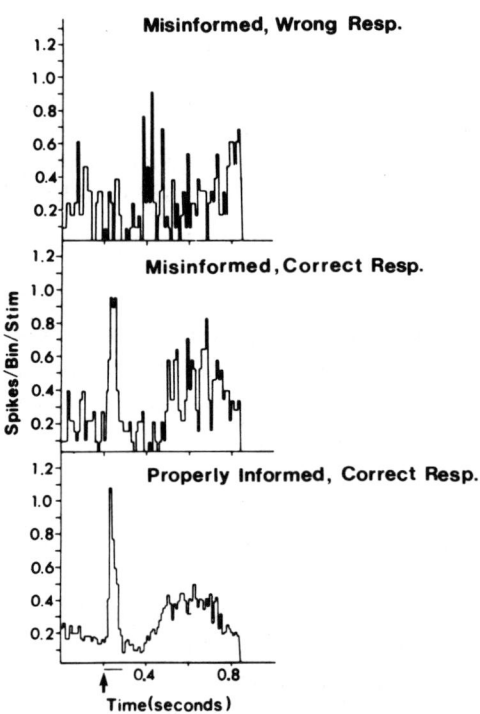

Figure 6.11. *Response of auditory cortex neuron to an auditory stimulus.* The stimulus onset is marked by the arrow. In the bottom histogram, a visual signal has informed the animal to expect that particular auditory stimulus. In the middle raster, he was misinformed, that is, he expected a different auditory stimulus, but performed the task correctly anyway. In the top histogram, the monkey was misinformed and performed the task incorrectly. Note that the clear responses to the stimuli seen in the middle and bottom histogram are not present in trials with erroneous responses. (From Hocherman et al., 1981, figure 8)

a facilitation of the transient auditory response when the visuospatial cue properly informed the animal and inhibition of that response when the cue misinformed; facilitation of the response in other neurons in trials when the monkey was misinformed. Interestingly, the auditory response was often totally absent in those trials in which the monkey made erroneous motor responses to the stimuli (Figure 6.11).

Sustained postinstruction activity has been observed by Vaadia et al. (1982) when the instructional significance of a stimulus depended on the monkey's immediately preceding behavior. In one block of trials, a noise stimulus instructed the monkey to make a leftward movement and a tone stimulus gave him instruction for a rightward movement. In another block of trials, the noise and tone stimuli instructions were reversed: the noise stimulus meant that the monkey should move his limb to the right and the tone stimulus meant to move left. Under these conditions, the response to the tonically applied auditory stimulus could be evaluated independently of the motor response that it instructed. As might be expected, within primary and adjacent auditory fields, most of the units responded to the stimulus in the same way, regardless of the instructional information it contained. In a number of neurons, however, the response was different for the two conditions. Some single-units responded to the tone only when it signaled a leftward movement. Vaadia et al. (1982) interpreted this specificity as indicating the existence of a stimulus-response association. These findings serve to emphasize once again the importance of motor set and instructions on the activity patterns of the cerebral cortex.

WHILE A MONKEY WAITS

In view of the studies discussed in this chapter, it appears likely that during the period of withholding response to an already presented instruction, sustained activity takes place throughout much of the cerebral cortex. This sustained activity may reflect or be the basis of perceptual and motor set. At the same time, perceptual and motor set have significant influences on the neuronal responses to sensory stimulation. It is not excessively speculative, we feel, to suggest that some of the sustained activity may be causally responsible for changes in neuronal input-output functions and the behavioral manifestations of set-related changes in behavior.

Chapter 7

Pathways for Set-Dependent Behavior

This consideration of pathways into and out of areas containing neurons that exhibit set-related activity focuses on the primary (or precentral) motor cortex (MI), the premotor cortex (PM), and the supplementary motor cortex (MII). Of these three areas, MI contains neurons whose activity is most clearly related to overt behavior. Indeed, the large pyramidal tract neurons of MI have been referred to historically as "upper" motoneurons by analogy with the "lower" motoneurons of the spinal cord. The remaining two areas (MII and PM) are intimately linked with MI, and the same behaviors that depend on set-dependent switching of MI input-output relations may also depend on set-dependent changes of tonic discharge in PM and MII. Thus, these three subdivisions of the frontal agranular cortex are sufficiently alike to allow them to be compared in the same behavioral paradigms, and yet are sufficiently different to suggest that they play different roles in motor control. We begin with the anatomy of some pathways that might be involved in set-dependent behavior, considering thalamocortical and corticocortical connections and contrasting the functional roles of these two classes of inputs to primary and nonprimary motor (i.e., the premotor and supplementary motor) areas. This is followed by consideration of some of the electrophysiological features of these pathways.

THALAMIC INPUTS TO FRONTAL AGRANULAR CORTEX

Figure 7.1 provides a historical background for current cytoarchitectonic definitions of the three zones (MI, PM, and MII) with which we are concerned. On the lateral surface of the hemisphere, MI corresponds approximately to Brodmann's area 4 and PM is a part of area 6. On the medial surface of the hemisphere, MII is also located in Brodmann's area 6. There have

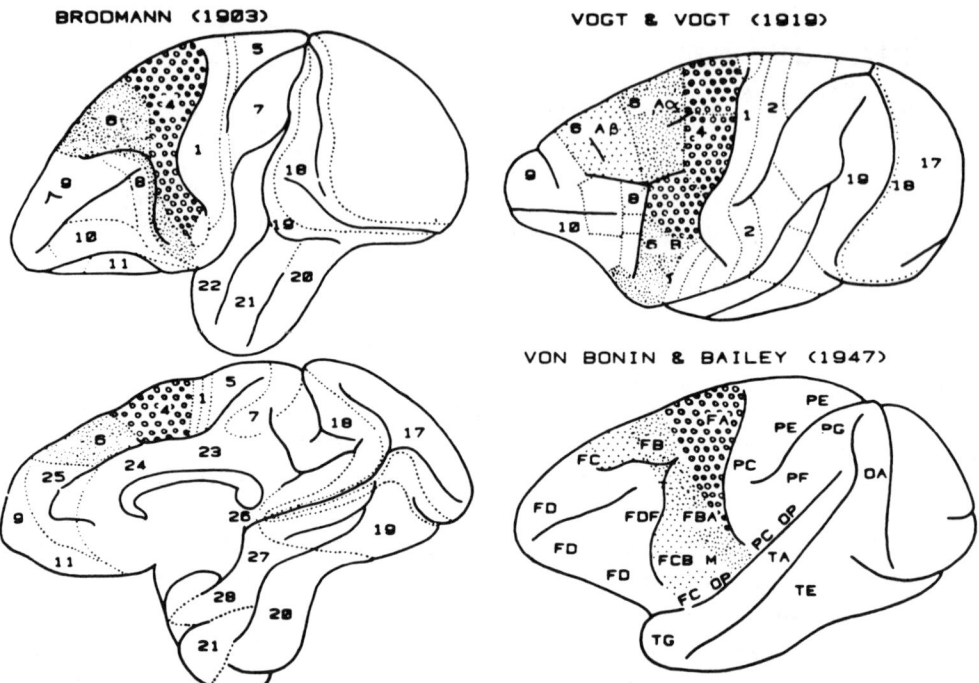

Figure 7.1. *Motor areas in relation to classic cytoarchitectonic maps of the monkey cortex.* Area 4 and FA correspond approximately to primary motor cortex. Note variations of anterior border in the three maps. Brodmann's area 6 (left maps) is subdivided by Vogt and Vogt and by von Bonin and Bailey. (From Wiesendanger, 1981)

always been disagreements as to the exact borders between the zones, whether the borders are set by electrical stimulation or by cytoarchitectonics (as in the maps of Vogt, Brodmann, or Von Bonin and Bailey shown in Figure 7.1). However, it is not our intention to attempt a reconciliation of these disagreements or of those aspects of projections from thalamus to cortex that remain controversial. Instead, we will dwell on the cortical maps and thalamic projection patterns on which there is general agreement, for these data will be most useful in formulating testable hypotheses as to pathways underlying set-related cortical activity.

The ventral group of thalamic nuclei project upon PM, MII, and MI. This group consists of the ventral anterior nuclei (VA), the ventral lateral (VL), divided into an oral part (VL_o), a medial part (VL_m) and caudal part (VL_c), and the oral part (VPL_o) of the ventral posterior lateral nucleus (see Jones and Porter, 1980; Asanuma et al., 1983). It is furthermore apparent that PM and MII receive inputs from more medial and rostral parts of the ventral thalamic group than does MI, which receives from its more lateral and caudal parts.

Strick (1976a) made horseradish peroxidase (HRP) injections into the arm area of MI, and observed a thin, continuous slab of labeled neurons in the ventrolateral thalamus flanked medially and laterally by unlabeled neurons. He also found that the leg representation of MI (located *medially* in cortex) is connected with the *lateral* part of ventral thalamus, whereas the face representation of MI (located *laterally* in cortex) is connected with the *medial* part of this thalamic group. Strick's observations on the topographical organization of the ventral thalamic arm area as defined by retrograde transport of HRP from MI (Figure 7.2) were consistent with his physiological observations, which showed that thalamic neurons related to active arm movements were situated in a restricted part of the ventral thalamic group; thalamic neurons related to spontaneous jaw and tongue movements were situated medially; and those related to spontaneous leg movements laterally (Strick, 1976b). This topographic pattern has been confirmed with both retrograde (Figure 7.3) and anterograde (Figure 7.4) fiber tracing methods.

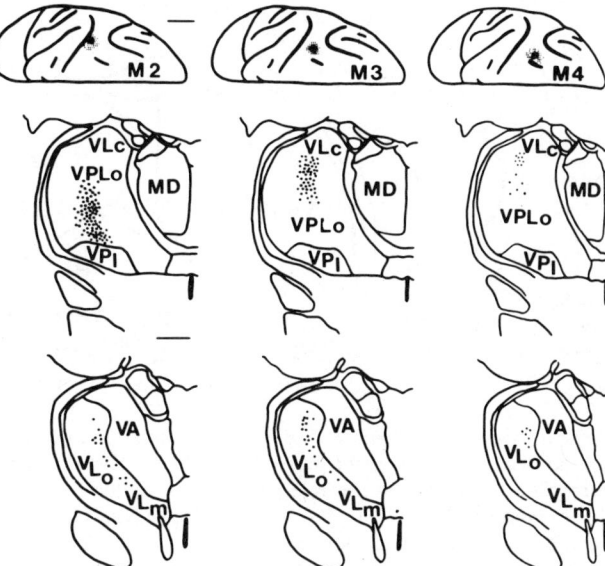

Figure 7.2. *Organization of the thalamic arm area.* Top row: a horseradish peroxidase injection site (black dot) and estimated spread of the injection (fine stipple) are indicated for three monkeys. Middle and bottom row: distribution of thalamic neurons (dots) labeled by retrograde transport following the cortical injections above each column. Note that labeled thalamic neurons occur more ventrally in the thalamic arm area as the injection site is placed nearer the central sulcus. (From Strick, 1976a)

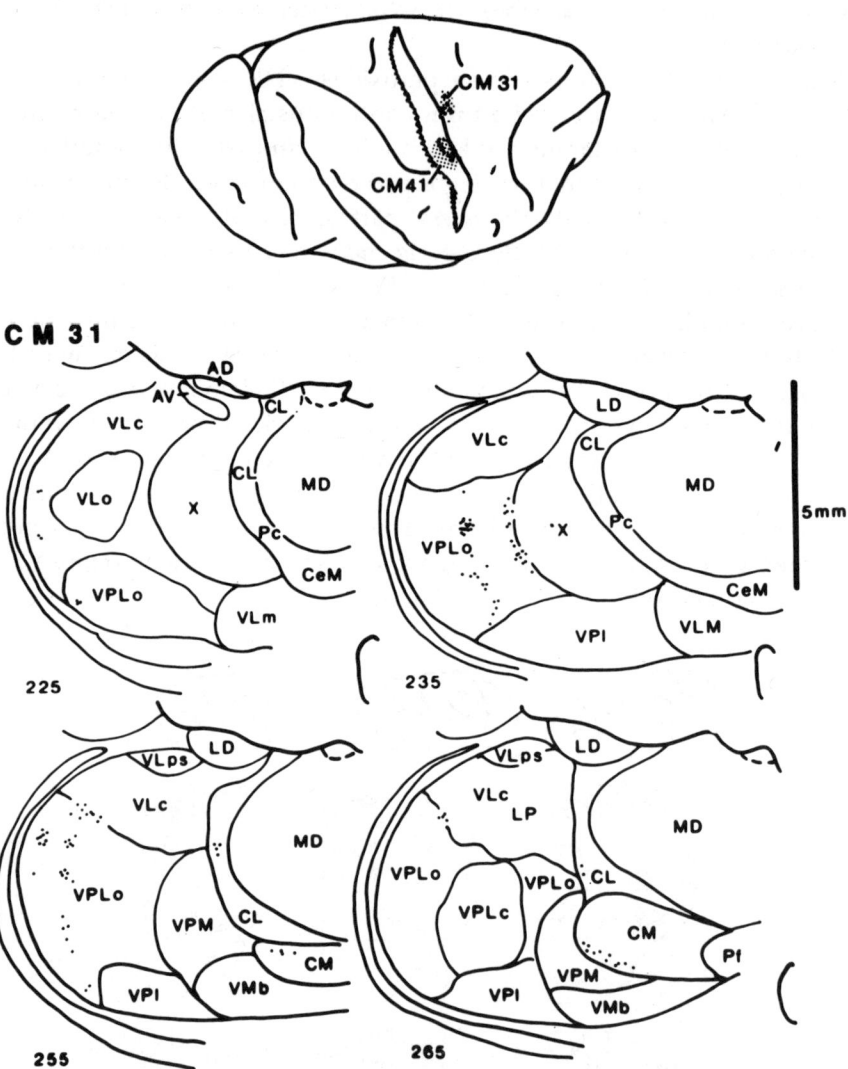

Figure 7.3. *Retrograde labeling of thalamic cells following a precentral peroxidase injection.* The top part of this figure shows injection sites of peroxidase (CM 31) and of tritiated amino acids (CM 41) that were used to determine the relation between motor cortex and thalamus by retrograde and orthograde transport (see Figure 7.4), respectively. In the diagrams of the thalamus, retrogradely labeled cells are shown in the part of the VL complex called VPL_o and, in addition, there are retrogradely labeled cells in two of the "diffusely projecting" thalamic nuclei, CL (centrolateral), and CM (centre median). (From Jones et al., 1979b)

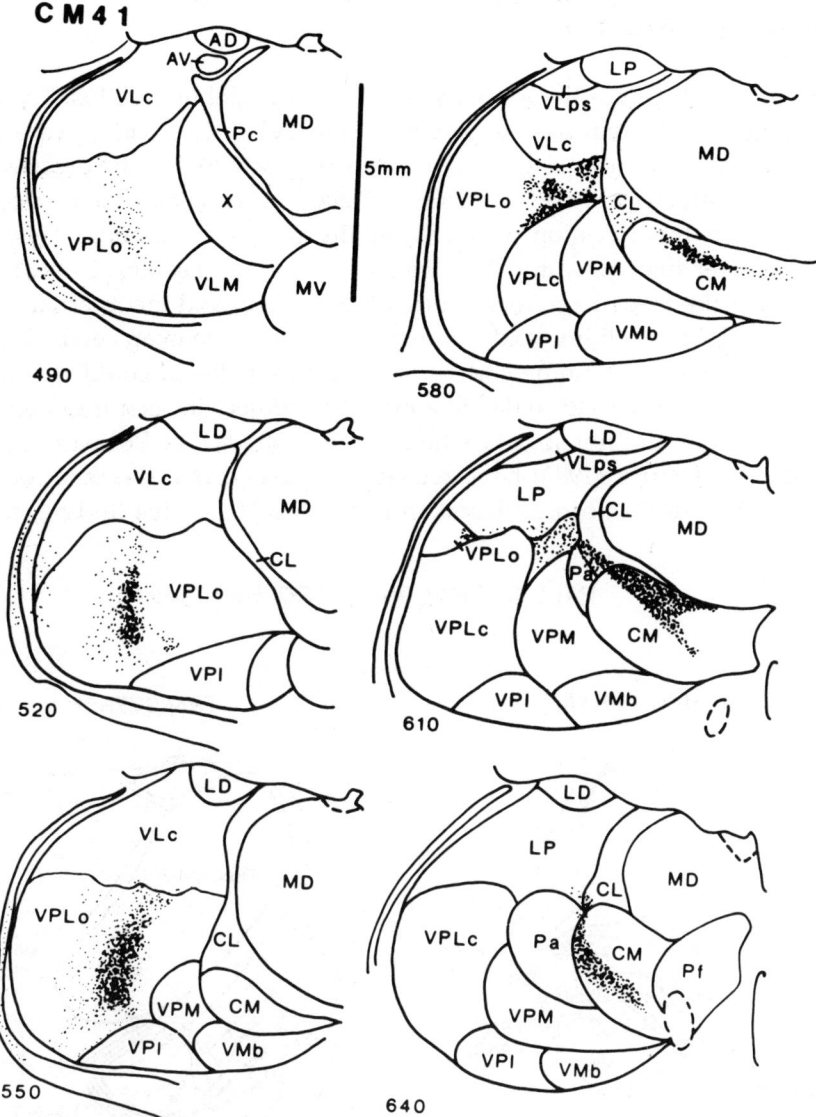

Figure 7.4. *Thalamic terminals of corticothalamic fibers.* An injection of tritiated amino acids (see Figure 7.3, CM 41) resulted in terminal labeling as shown by the stippling. In addition to the dense labeling in VPL$_o$, there was also labeling in two diffusely projecting thalamic nuclei (CL and CM). Abbreviations as in Figure 7.3. (From Jones et al., 1979b)

INPUTS TO THALAMUS FROM GLOBUS PALLIDUS AND CEREBELLUM

Figure 7.5 diagrams the output pathways of the globus pallidus. The ventral thalamic nuclei that receive pallidal inputs (VL_o, VL_m, and VA) are known to project to the frontal agranular cortex (Figure 7.6), but the precise targets of these nuclei are still uncertain. Tracey et al. (1980) have specifically examined this question with retrograde transport methods. They found that tracer injections into VL_o lead to retrograde labeling in both rostral parts of the frontal agranular cortex (i.e., area 6) and the internal segment of the globus pallidus (and, importantly, not in the deep cerebellar nuclei or the precentral motor cortex). The globus pallidus could therefore be thought to project to thalamic zones that project in turn into area 6, but Tracey et al. did not specify whether it was the MII or PM parts of area 6. Schell and Strick (1983) have recently suggested that the main output of the pallido-thalamocortical system is MII (via VL_o). This finding, together

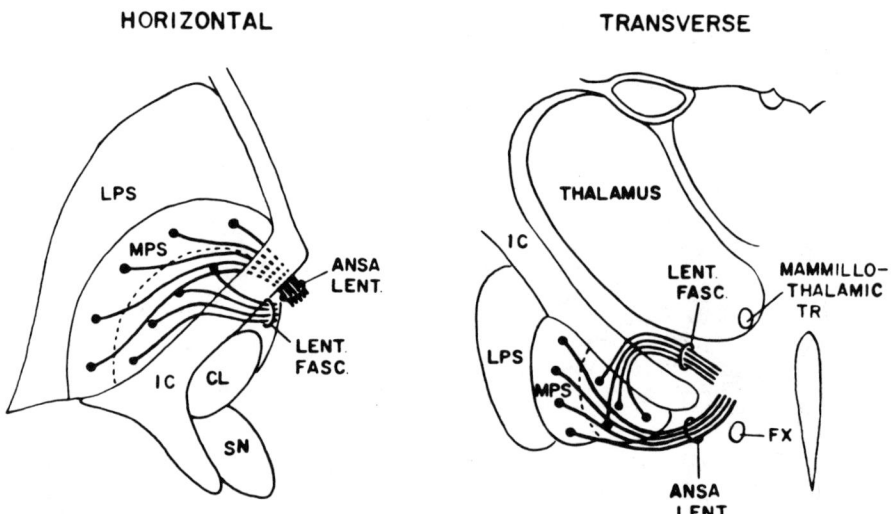

Figure 7.5. *Fiber bundles linking globus pallidus with thalamus.* The diagrams show the origin and course of fibers forming the ansa lenticularis (ansa lent.) and lenticular fasciculus (lent. fasc.) in the monkey. Fibers of ansa lenticularis arise from the outer portions of the medial (or internal) pallidal segment (MPS), and fibers forming lenticular fasciculus arise from inner parts of the medial pallidal segment. Abbreviations: CL, subthalamic nucleus; FX, fornix; IC, internal capsule; LPS, lateral (or external) pallidal segment; SN, substantia nigra. (From Kuo and Carpenter, 1973)

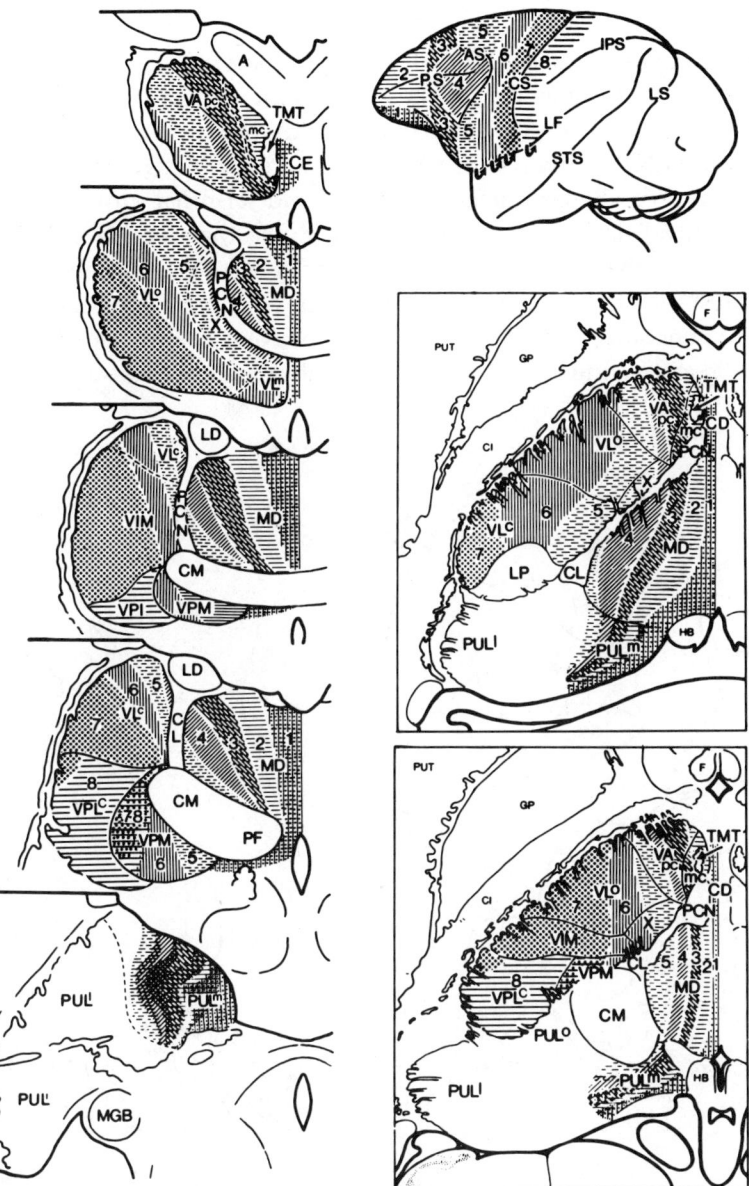

Figure 7.6. *General pattern of thalamocortical projections.* The general pattern of connectivity between thalamus and cortex is shown both in frontal (left) and horizontal sections (right). It is apparent that the more rostral regions of cerebral cortex are linked with the more medial zones of thalamus, and that more caudal regions, such as the postcentral gyrus (strip 8), receive inputs from caudal and lateral thalamus, that is, VPL$_c$ (shown as zone 8 in the diagrams of thalamus). Nonprimary motor cortex (strip 5 and part of strip 6) receives inputs from thalamic areas referred to as VA, VL$_o$, and VL$_c$. Finally, primary motor cortex (strip 7) receives from a zone of thalamus generally intermediate between that projecting to the postcentral gyrus and that projecting more rostrally to nonprimary motor cortex. (From Kievit and Kuypers, 1977)

with the existence of set-related activity in striatal neurons (Rolls et al., 1983) and MII (Tanji et al., 1980), points to the importance of investigating a cortex-to-striatum-to-globus pallidus-to-thalamus-to-cortex circuit as we seek to identify pathways for set-related activity in cerebral cortex.

The deep cerebellar nuclei appear to send terminal projections to parts of the ventral tier group different from those that receive pallidal input. Figure 7.7 shows retrogradely labeled cells in the deep cerebellar nuclei after HRP, a retrograde axonal marker, has been injected into VL$_c$. Such an HRP injection produces retrograde transport in two directions (to cortex via corticocothalamic fibers and to cerebellum via cerebellothalamic fibers),

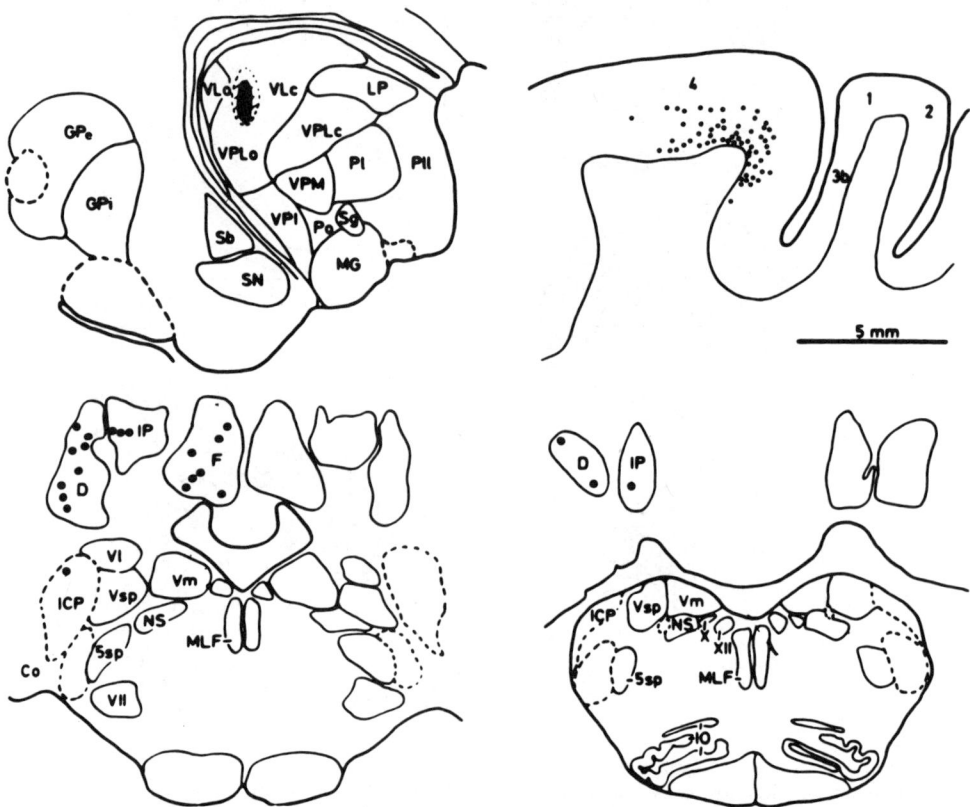

Figure 7.7. *Sources of inputs to a thalamic relay nucleus.* An injection of peroxidase into VL$_c$ (upper left) resulted in retrograde labeling in the dentate nucleus (D), interpositus nucleus (IP), and fastigial nucleus (F). The injection in VL$_c$ failed to cause retrograde labeling in globus pallidus (not shown). That the injection site at upper left was indeed within the zone projecting to motor cortex is indicated by the results at upper right, depicting corticothalamic cells in primary motor cortex that also were labeled by retrograde transport. (From Tracey et al., 1980)

and so has the capacity to identify the loci within cerebellar nuclei projecting via thalamus to a given zone of cortex. Cortical labeling is observed in MI and the cerebellar labeling in all three cerebellar nuclei. The VL_c injection was not followed by labeling in the globus pallidus, nor was labeling found in the dorsal column nuclei.

In broad outline, then, it appears that the zones of thalamus which project to MI receive inputs from cerebellar nuclei, whereas parts of thalamus that project to MII receive inputs from the globus pallidus. Schell and Strick (1983) report that the input to PM is also from the cerebellum. The cerebellar inputs to thalamus arise primarily from interpositus and dentate nuclei, with weaker projections from the fastigial nucleus (Asanuma et al., 1983).

CORTICOCORTICAL INPUTS: PARIETAL CORTEX

Even before Sperry's discoveries on the effects of disconnecting the hemispheres by surgical interruption of the corpus callosum, a number of studies had shown "disconnection syndromes" in patients with neurological disorders (see Geschwind, 1965 for the early literature). However, only after Sperry's work did studies in nonhuman primates begin to reveal the deficits resulting from the interruption of ipsilateral corticocortical connections (see Haaxma and Kuypers, 1975). There are projections to MI from a number of the cortical areas, and it is convenient to divide them into two groups, according to whether they arise from the parietal lobe or from other parts of the frontal lobe.

Figure 7.8 (Strick and Kim, 1978) shows results of two experiments in which cells in area 5 were labeled by retrograde transport of HRP injected into MI. These observations indicate that area 5, the rostral part of the posterior parietal cortex, has direct access to the arm area of the motor cortex. Electrophysiological confirmation of the anatomical fiber-tracing results of Strick and Kim was provided by Zarzecki et al. (1978b). As shown in Figure 7.9, MI was stimulated and antidromically activated units were recorded in area 5. Additional data on the projection from area 5 to MI has been provided by Jones et al. (1978) in experiments demonstrating the terminal distribution in MI after radioactive amino acids were injected in area 5, and showing that, in addition to its projection to MI, parts of area 5 also project to PM and MII. Areas 1 and 2 and the second somatic sensory cortex also project to MI, but another parietal somatic sensory field, area 3b, apparently does not (Jones et al., 1978; Vogt and Pandya, 1978). Although area 3b does not project forward into MI, it has dense projections caudally into area 1 and, to a lesser extent, into area 2, regions that do project directly

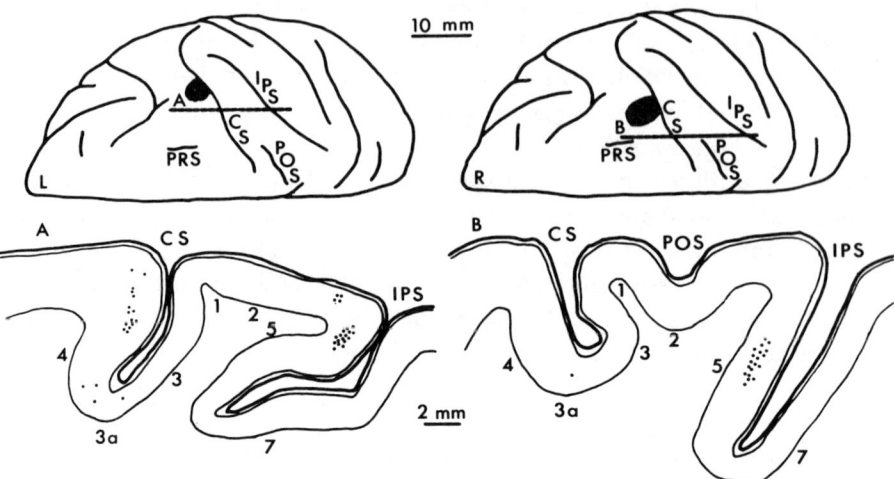

Figure 7.8. *Corticocortical inputs to MI from area 5.* Top: loci of two peroxidase injections in primary motor cortex that retrogradely label corticocortical cells projecting into MI. Bottom: labeled neurons (dots) found in representative parasagittal sections showing that area 5 contained the greatest number of labeled neurons. Abbreviations: CS, central sulcus; IPS, intraparietal sulcus, POS, postcentral sulcus; PRS, superior precentral sulcus. (From Strick and Kim, 1978)

to MI. Thus the somatosensory information that reaches area 3b (skin mechanoreceptor information) must be further processed in areas 1, 2, and/or 5, and cannot directly influence activity in MI (see also Hyvärinen and Poranen, 1978a, b)

Within the parietal lobe of the monkey, the fundus of the intraparietal sulcus is a dividing line between two areas (Brodmann's areas 5 and 7) with very different functional properties. Cells in areas posterior to the fundus of the sulcus are related to goal-directed ocular fixation or are responsive to visual stimuli (Mountcastle et al., 1975). Differences in the corticocortical projections from areas 5 and 7 parallel these functional differences. Whereas area 5 projects directly to MI, projections from area 7 (the inferior parietal lobule) bypass MI to terminate in the general region of PM (Pandya and Kuypers, 1969; Chavis and Pandya, 1976). Indeed, MI is strongly interconnected with other fields of the somatic sensorimotor cortex, but not directly interconnected with the visual or auditory areas of the temporal, parietal, or occipital lobes. Information from these visual areas can reach MI only indirectly, either via multisynaptic, cortico-subcorticocortical pathways, or connections from PM (Pandya and Kuypers, 1969; Jones and Powell, 1969, 1970; Chavis and Pandya, 1976).

We have now considered some of the pathways over which signals might be transmitted to interact in MI. The signals might pass to MI over two or more parallel channels. For example, signals from parietal cortex (area 7) to premotor cortex might traverse both a corticocortical route and a cortico-subcorticocortical route. The latter might involve both cortico-striato-pallido-thalamo-corticocortical and cortico-ponto-cerebello-thalamocortical routes. The cortico-subcorticocortical signals indicated systematically in Figure 7.10 would arise from output cells with somata in layer V of the cortex; the corticocortical connections would arise primarily (but by no means exclusively) from cells with somata in the supragranular cortical layers. By virtue of the different cell classes from which they arise and the different pathways over which they are relayed, there is reason to believe that corticocortical and thalamocortical inputs to MI might transmit different sorts of signals (see Chapter 5). The patterns of thalamocortical and corticocortical terminations in cortex may provide clues as to their different functions. In a

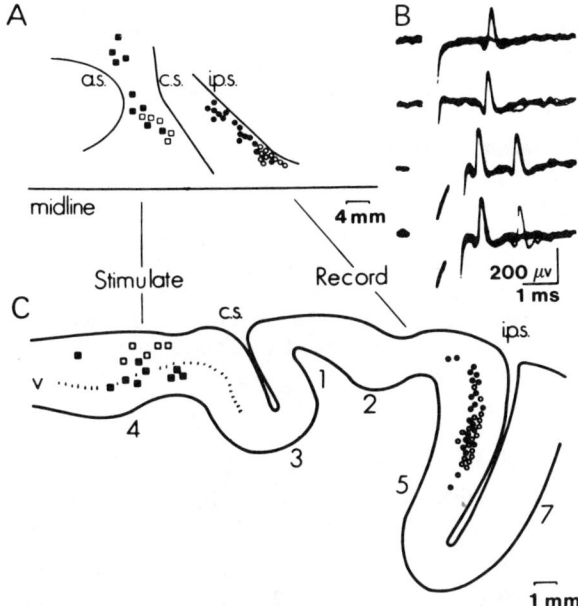

Figure 7.9. *The projection of area 5 to primary motor cortex.* A: the surface locations of penetrations for microstimulation (squares) and single-unit recording (circles). Open squares: penetrations where intracortical stimulation evoked muscle contraction at wrist or elbow. Open circles: penetrations in which neurons were activated from sites marked by open squares. B: identification of antidromic responses. C: summary of locations of effective stimulating sites (squares) and antidromically invaded neurons (circles) on a standard parasagittal section. Abbreviations: v, layer five of cerebral cortex; a.s., arcuate sulcus; c.s., central sulcus; i.p.s., intraparietal sulcus. (From Zarzecki et al., 1978b)

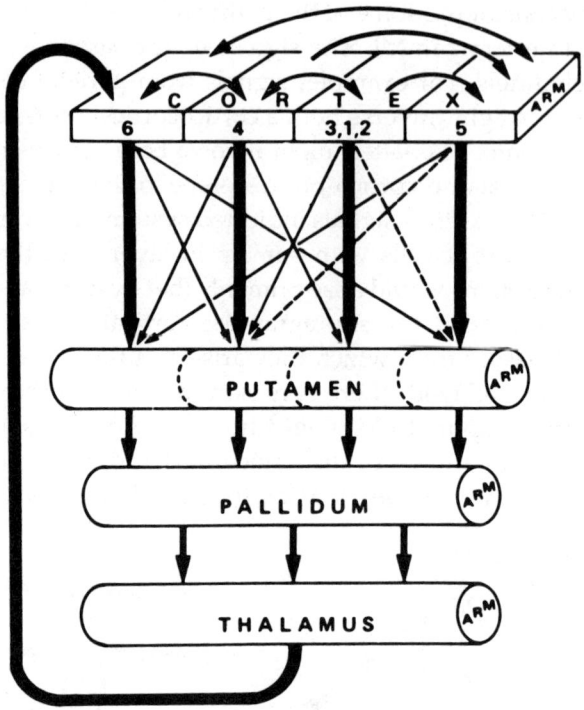

Figure 7.10. *Some cortical sources of inputs to nonprimary motor cortex.* Area 6, including both a lateral part (premotor cortex) and a medial part (MII), receives inputs from other parts of the sensorimotor cortex either via corticocortical connections (indicated by arrows at the top of this diagram) or via the putamen, globus pallidus, and thalamus, as shown by the successive vertical arrows. In addition to receiving inputs from the putamen, the globus pallidus also receives inputs from the caudate nucleus, which in turn receives inputs from prefrontal cortex. Some of these circuits might play a role in set-dependent neuronal switching processes. (From DeLong and Georgopoulos, 1981)

review, Jones (1981a) noted that the corticocortical fibers to any cortical area terminate in small focal zones, separated by zones with rather sparse inputs. However, this focal termination does *not* mean that there are interdigitating foci of cortical cells in which some foci are influenced and other foci are uninfluenced by corticocortical inputs. For example, there is no information that rules out a very high efficacy for corticocortical synapses in the relatively sparse zones. Indeed, as we examine the details of both thalamocortical and corticocortical terminal distribution, we are confronted by the limits of functional inferences on the basis of relative densities of terminals. Information as to the terminal distribution of corticocortical inputs fails to tell us about the possibilities for neuronal interaction within the cortex, so it is essential to supplement anatomical with electrophysiological data.

ELECTROPHYSIOLOGICAL ANALYSES OF INTERCONNECTIONS

The *absence* of a direct pathway (as revealed by HRP or radioactive protein transport) between two areas rules out a direct, monosynaptic effect of one area on the other, but does not rule out multisynaptic effects. Conversely, the *presence* of a pathway does not reveal the strength or the sign (inhibition or excitation) of the effect that the pathway mediates. It is thus apparent that although anatomical data provide us with a list of areas that have the potential for playing a role in set-related input-output switching, these data do not enable us to identify the effects on those areas. Electrophysiological observations should reveal dynamic features of the anatomically defined pathways and establish their role in set-related brain activity and behavior.

Our discussion of electrophysiology parallels that of anatomy in focusing on corticocortical and thalamocortical inputs to area 4 and area 6. Ideally, a review of the thalamocortical inputs would be subdivided according to "specific" inputs, such as those from the cerebellum and the globus pallidus to the frontal cortex via the main relay nuclei of the thalamus, and putatively "nonspecific" inputs, such as those to the frontal agranular cortex from the intralaminar nuclei of the thalamus, including the centre median nucleus and the central lateral nucleus (see Strick, 1976a; Jones, 1981b). For many reasons, however, most is known about the specific inputs from cerebellum, with less extensive knowledge of those from the globus pallidus. As a result, our review deals primarily with the cerebello-thalamocortical circuit, rather briefly with observations on pallido-thalamocortical electrophysiology, and even less with nonspecific inputs. The relative lack of useful electrophysiological analyses of the pallido-thalamocortical and nonspecific systems is unfortunate, as both may be highly important in set-related functions.

CEREBELLAR PATHWAYS TO FRONTAL AGRANULAR CORTEX

Whereas the anatomical work that was presented in the first part of this chapter was derived in large part from the monkey, the electrophysiological data that we will now consider were obtained in the cat. This is unfortunate, because it means that any conclusions must be considered only tentative for our purposes, but is necessary, because all of the postulated experiments (see Chapters 8 and 9) involve monkeys. In this discussion of the cerebello-thalamocortical pathway in cats, VL is used to refer to a complex of subnuclei thought to be homologous to VPL_o and VL_c in certain monkeys.

As shown schematically in Figure 7.11, fibers emerging from the interpositus (IP) and the dentate (DEN) nuclei of the cerebellum terminate on

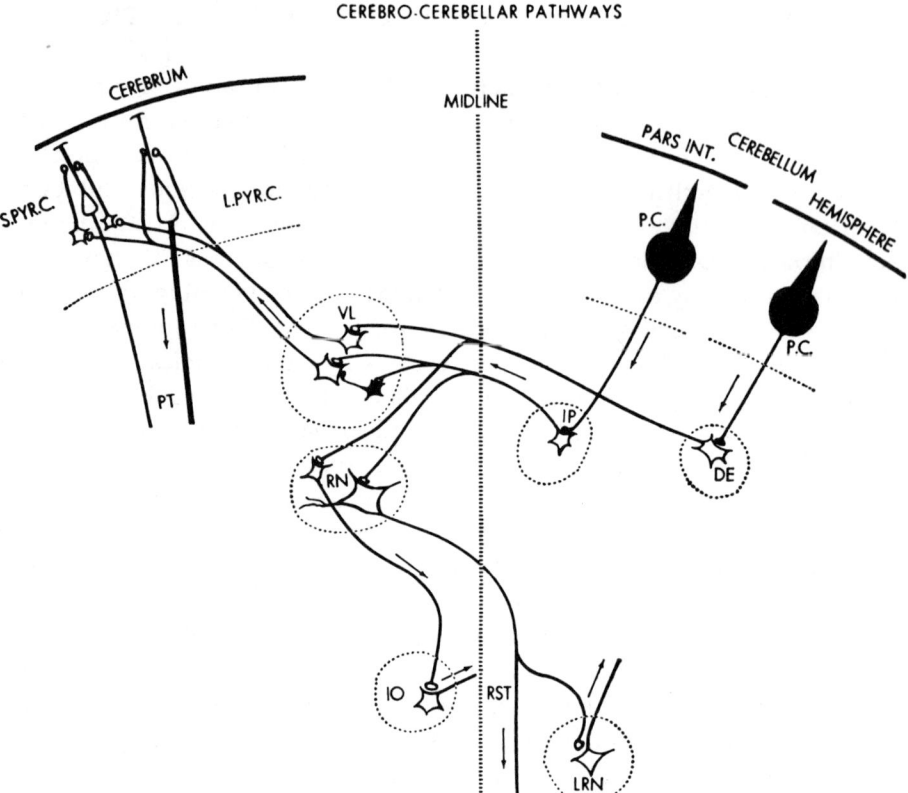

Figure 7.11. *Pathways from cerebellum to cerebral cortex via the thalamus.* Two of the cerebellar output nuclei, interpositus (IP) and dentate (DE), project to the VL complex of the thalamus. Second-order neurons with cell bodies in the thalamus project to cerebral cortex, where they have strong excitatory effects on large pyramidal tract cells (L.PYR.C.) and weaker, mainly disynaptic excitatory effects on small pyramidal tract cells (S.PYR.C.). Abbreviations: PT, pyramidal tract; RN, red nucleus; IO, inferior olive; LRN, lateral reticular nucleus; PARS INT., pars intermedia; P.C., Purkinje cell; RST, rubrospinal tract. (From Allen and Tsukahara, 1974)

cells in VL; these cells, in turn, project to motor cortex (Allen and Tsukahara, 1974). It is also shown that the larger MI PTNs receive direct *mono*synaptic inputs from VL (and therefore *di*synaptic inputs from cerebellum), whereas smaller MI PTNs receive relatively few monosynaptic inputs from VL (cf. Deschenes et al., 1982). Cortical interneurons are interposed between most VL terminals and the smaller PTNs, which are mainly excited *di*synaptically from VL (and therefore *tri*synaptically from the cerebellum).

The fine-structure of the synapse (as revealed by the electron microscope) is a potential clue about the sign of synaptic action, but most current knowledge as to whether a given synapse is excitatory or inhibitory has been provided by electrophysiology. Electrophysiological data for the synapses between cerebellothalamic terminals and VL cells were provided by Uno et al. (1970). Figure 7.12 from their study shows that inputs from cerebellum to VL cells are excitatory, since EPSPs were recorded with monosynaptic latency in VL neurons in response to cerebellar nuclear stimulation.

One of the major areas with which we will deal in Chapter 9 concerns shifts in dominant control of MI neurons between the dentate nucleus (DEN) and interpositus nucleus (IP) depending on motor set, and experiments to demonstrate these shifts utilize the recording of VL neuronal activity evoked by DEN and IP stimulation during different motor sets. Therefore, it is important to consider the interaction of these two pathways at the thalamic and cortical levels. Figure 7.13 diagrams the several steps used by Shinoda et al. (1982) to compare the effects of IP and DEN on VL neurons that project to MI. In these studies, EPSPs evoked by cerebellar nuclear stimulation were recorded from VL neurons that were antidromically activated

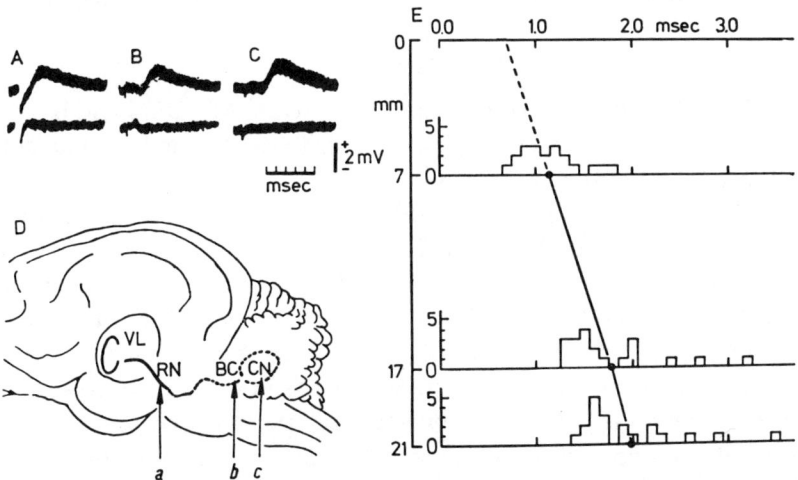

Figure 7.12. *Monosynaptic excitation of VL neurons by cerebellar output fibers.* A, B, and C: EPSPs evoked in a VL relay neuron by electrical stimulation at three different sites within the pathway from cerebellum to VL thalamus. D: The closest site (A) was in the red nucleus (RN), where cerebellar fibers passing through and giving off collaterals may be directly activated. The second site of stimulation (B) was the brachium conjunctivum (BC), where cerebellar output fibers may be activated as a bundle. The farthest point of electrical stimulation was at (C), where the cerebellar nuclei (CN) may be directly stimulated. E: The differences in latencies for a series of EPSPs recorded in VL neurons are shown at the right. It is apparent that there is a linear change in EPSP latency as the stimulus site is moved further away from VL. There are no abrupt jumps in latency, as would be produced by a synaptic delay intervening between the cerebellar output and the EPSP of the VL neuron. This points to a monosynaptic EPSP in the VL neuron as a result of activation of the cerebellar output. (From Uno et al., 1970)

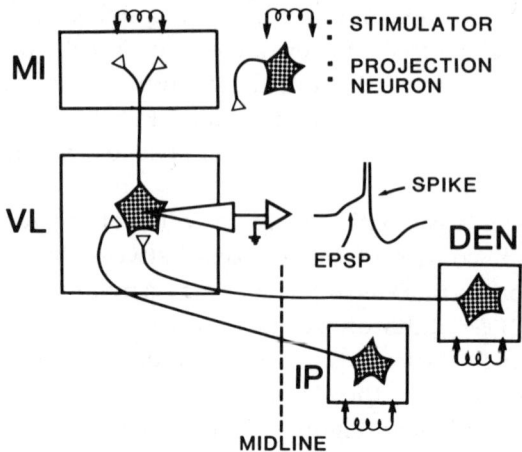

Figure 7.13. *Schematic representation of a set-up to compare effects of interpositus (IP) and dentate (DEN) on VL neurons that project to MI.* Three aspects of the set-up are most important: a microelectrode is inserted into a VL neuron to record EPSPs evoked by DEN and/or IP stimulation; a stimulus is delivered to MI to see if the VL neuron can be activated antidromically and thereby identified as a cell that projects to MI; and DEN and IP stimuli are given separately or in a variety of temporal sequences to determine their interactions at the level of individual VL neurons.

by stimulation of MI to show that they projected to MI. DEN and IP stimuli were given in a variety of temporal sequences to determine their interactions at the level of individual VL neurons.

Figure 7.14 illustrates EPSPs recorded from a VL neuron in response to electrical stimulation in DEN and IP, and supports the view that there is convergence onto this VL neuron from both of these deep cerebellar nuclei. The figure shows that the same VL neuron exhibits four different "unitary" EPSPs evoked by stimuli at four different sites in the cerebellar nuclei. The conclusion that these EPSPs are unitary is based on the observation that they are "all-or-none." Thus, in the sets of superimposed traces in Figure 7.14B, EPSPs are sometimes present and sometimes absent, but there are no EPSPs with gradually augmenting amplitudes.

The presence or absence of unitary EPSPs is the result of the stimulus intensity being adjusted to "straddle" the threshold of the particular neuron in DEN or IP whose synapse in VL was responsible for the unitary EPSP recorded from the VL neuron. If it can be established that the effects of the electrical stimuli do not spread beyond the nuclei in which the stimulating electrodes or the fibers are located or to fibers from other cerebellar nuclei passing by or through the stimulated nuclei (see Shinoda et al., 1982, 1983, for discussion), the data mean that an individual neuron in VL may receive synapses from two different cerebellar nuclei—that is, outputs of DEN and IP converge on individual VL neurons (Shinoda et al., 1982, 1983). However, not all VL neurons receive such convergent input, and it remains possible that the proportion of cell receiving from both DEN and IP is lower than that estimated by Shinoda et al. (1982). This lower estimate could result from unavoidable stimulation of fibers from the dentate nucleus as they pass through the interpositus, especially in the rostral part of the interpositus nucleus.

Additional features of the experimental techniques used to investigate the way in which DEN and IP inputs interact in VL are shown in Figure 7.15, where progressively increasing stimulus intensities in DEN and IP first evoked a single unitary EPSP and then the summation of this first EPSP combined with a second unitary EPSP.

Figures 7.14 and 7.15 illustrate the sorts of data that allow one to suggest that some DEN and IP signals may converge on individual VL neurons. But to get some idea as to the proportion of VL neurons that receive inputs

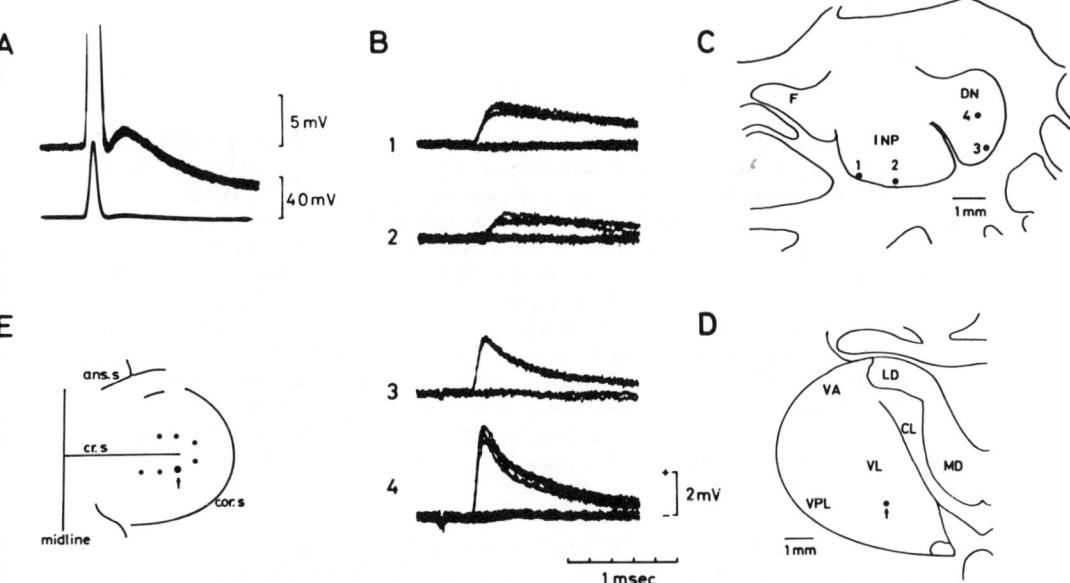

Figure 7.14. *Convergence of dentate and interpositus projections onto thalamocortical neurons in VL.* A: Antidromic responses of a thalamocortical neuron evoked by stimulation of the motor cortex. B: Monosynaptic EPSPs evoked in an all-or-none fashion from the IP (1, 2) and DN (3, 4) with less than 75 microamperes. C: Histological identification of the stimulating sites in the cerebellar nuclei. D: Recording site of this thalamocortical neuron in VL. E: Localization of cortical site from which the cell could be antidromically activated. Abbreviations: ans. s, ansate sulcus; cr. s, cruciate sulcus; cor. s, coronal sulcus. (From Shinoda et al., 1982)

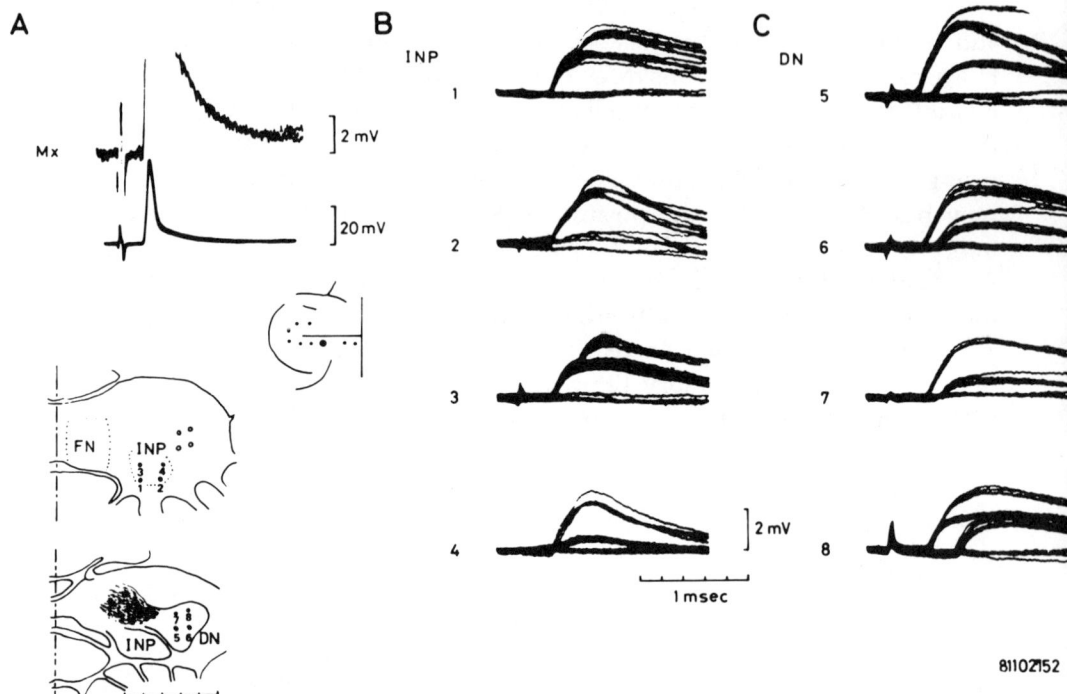

Figure 7.15. *Unitary VL EPSPs evoked by stimulation of interpositus and dentate.* Column A: The VL neuron's antidromic response to stimulation of MI, establishing that the neuron from which EPSPs were recorded was a projection neuron. Indicated at the lower left are the locations of four stimulating-site electrodes in the posterior part of the interpositus (INP) nucleus (dots marked 1, 2, 3, and 4) and four additional stimulating sites in the dentate (DN) nucleus (marked 5, 6, 7, and 8). Column B: VL EPSPs, the numbers indicate the stimulating sites in the interpositus nucleus that evoked the EPSPs. Column C: VL EPSPs, the numbers indicate the stimulating sites in the dentate nucleus. For all stimulating sites, a gradual increase of stimulus intensity from below the threshold produced multiple EPSPs in an all-or-none manner. These EPSPs had different shapes, time courses, and latencies. Therefore it was apparent that individual unitary EPSPs were evoked by stimulation of different single neurons or fibers within the cerebellar nuclei. This, in turn, points to a degree of convergence of inputs from the cerebellar nuclei to the same VL neuron. (From Shinoda, unpublished)

from both DEN and IP and the relative strengths of these inputs to neurons in different parts of VL, it is necessary to record systematically from many VL neurons. Such exploration indicates that some 40% of VL neurons appear to receive inputs from both DEN and IP, whereas the others receive inputs from only one of these two cerebellar nuclei (Figure 7.16). For the reasons stated above, as well as others, the exact proportion of VL cells that receive convergent inputs from DEN and IP is difficult to determine, but it is unlikely to be either all or none.

The functional significance of these inputs, both convergent and non-convergent, can be appreciated by reviewing some ideas on the role of the intermediate zones of cerebellar cortex, which projects primarily via IP, as contrasted to the role of the cerebellar hemispheres, which project primarily via DEN. Allen and Tsukahara (1974) outlined a scheme in which the cerebellar cortex pars intermedia (via IP) interacts with motor cortex pyramidal tract neurons during movement. It was noted that, simultaneously with the pyramidal tract discharge to the spinal motoneurons, the pars intermedia

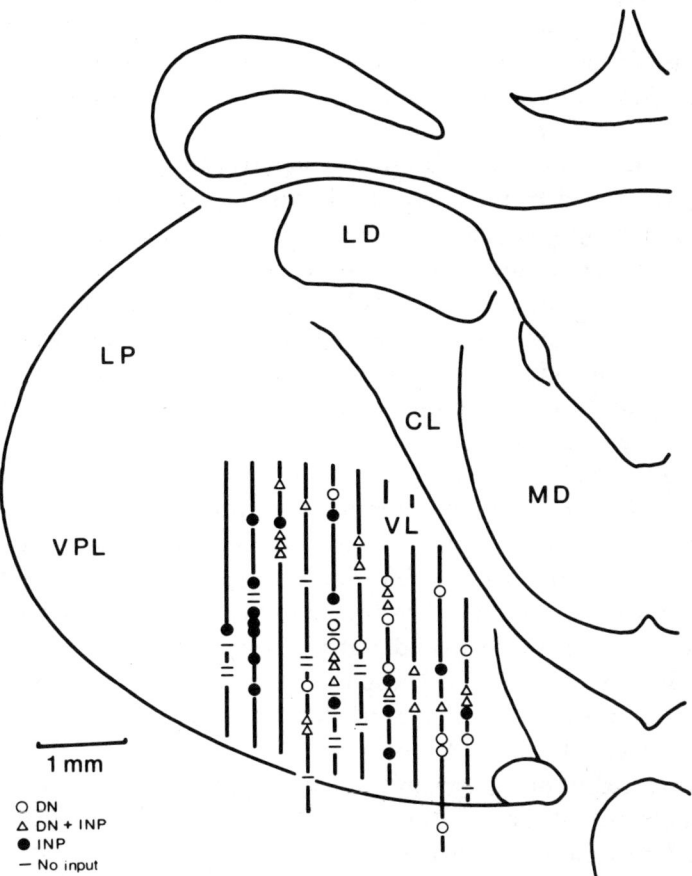

Figure 7.16. *Distribution of interpositus (INP) and dentate (DN) inputs to thalamocortical neurons in VL.* A single micropipette was used throughout this experiment and 73 cells were penetrated intracellularly. Thalamocortical neurons were identified as such by their antidromic responses to stimulation of the primary motor cortex and area 6. Open circles, dark circles, and triangles indicate neurons receiving inputs from the dentate, interpositus, and both, respectively. Dashes indicate neurons which did not receive any apparent inputs from either nucleus. (From Shinoda, unpublished)

of the cerebellum receives indirect inputs from pyramidal tract collaterals. Then, with onset of movement, cutaneous, muscle, and joint receptors also send signals to the same zones of the cerebellum (by way of the spinocerebellar tracts) that have received the corollary discharge from pyramidal tract collaterals. Of course, after movement, the cortical inputs to the cerebellum would arrive earlier than the spinal inputs, and the Purkinje cell output from cerebellar cortex would therefore depend both on the pyramidal tract signal representing the motor command and a spinal input that provides a feedback signal. Allen and Tsukahara suggested that the pars intermedia cooperates with the cerebral cortex in feedback control of movement, and it followed that the pars intermedia would not perform corrections of movements by itself unless the cerebral cortex were functioning properly. These investigators suggested that, unlike the pars intermedia, the lateral cerebellar hemisphere integrates inputs primarily from sensorimotor cortex as a whole and returns signals primarily to the motor area.

Thus far, our electrophysiological discussion has been devoted primarily to the monosynaptic effects of cerebellar stimulation as recorded in thalamic neurons. But useful information can also be obtained by determining the multisynaptic effects of such stimulation on cortical neurons. Figure 7.17 illustrates some of the features of experiments in which electrical stimulation of the brachium conjunctivum (BC) was used to investigate MI responses to cerebello-thalamocortical inputs. A well-placed pair of electrodes in BC can activate a large part of the ascending cerebellar output, but in order to find the best point for stimulation within BC, it was necessary to measure the amplitude of the MI field potential as a function of the location of the stimulating electrode. To obtain such data, two stimulus intensities were used: 400 microamperes and 200 microamperes. In experiments such as this, it is important to stimulate the *entire* BC, for if only a portion of the BC were stimulated one would not know whether the lack of EPSPs in certain MI cells was because these cells did not have any inputs from the cerebellum, or was merely because the BC stimulus had failed to excite some BC fibers. In order to maximize the possibility of placing the BC stimulating electrode in the maximally effective locus (as inferred from cortical field potential amplitude), the 200 microampere stimulating current was used to locate the most effective point in the BC; a supramaximal stimulus intensity (500 microamperes) then was used to activate the entire BC. By using a BC stimulating electrode placed in the manner shown in Figure 7.17, it is possible to determine the effects of cerebellar signals on pyramidal tract neurons identified by their antidromic responses to pyramidal tract stimulation. The latency of the antidromic response provides information as to the axonal conduction velocity of the PTN, and allows the PTN to be categorized as fast, slow, or intermediate.

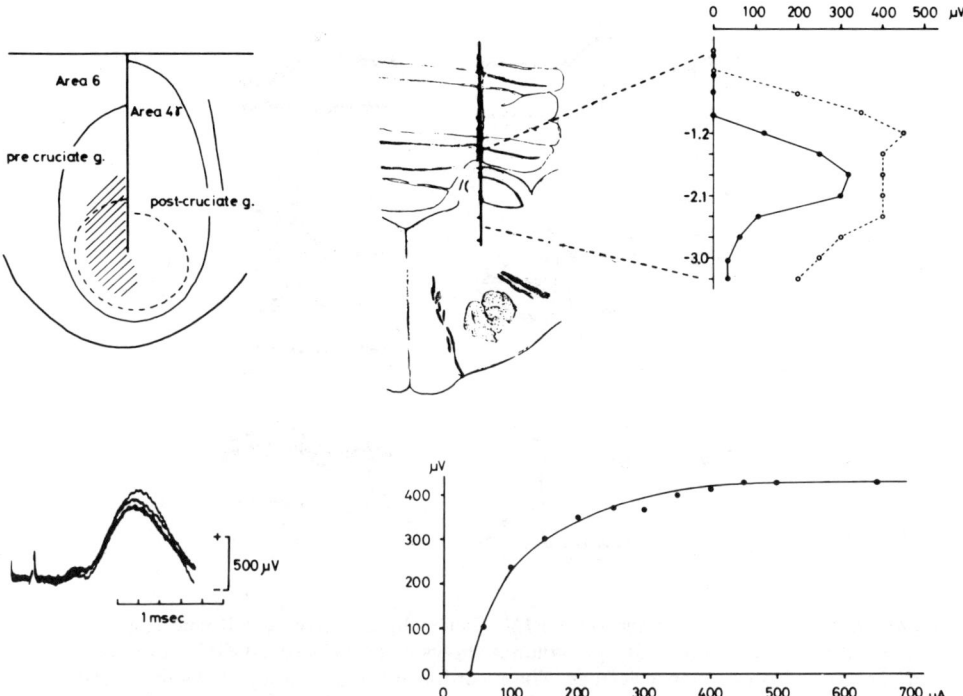

Figure 7.17. *Brachium conjunctivum stimulation.* An electrode in the brachium conjunctivum (BC) can stimulate a large part of the output from the interpositus and the dentate nuclei, and evoke cortical field potentials. Upper left: dotted line indicates the forelimb area of the cat motor cortex (surface view: medial is up, rostral is left) and the hatched area indicates the area from which the largest surface-positive potentials could be evoked by stimulation of the contralateral BC. Lower left: typical cortical potentials evoked from the BC. Upper right: stimulating electrode position and its relation to response amplitude of cortical potential. Stimuli were delivered at 300 micrometer intervals at intensities of 200 microamperes (closed circles, solid line) and 400 microamperes (open circles, dashed line). The amplitudes of the cortical field potentials recorded on the same point of the lateral precruciate gyrus (ordinate) are plotted against the depths of the electrode tip (abscissa). The most effective stimulating site (at a depth of 1.9 mm) corresponds to the position of the BC. Lower right: the relation between stimulus intensities and amplitudes of cortical potentials. (From Shinoda, unpublished)

Figure 7.18 shows the cerebellar-evoked EPSPs in PTNs with different axonal conduction velocities. In fast PTNs (conduction velocities greater than 40 meters per second), stimulation of BC produced large amplitude EPSPs with short latencies; these commonly exceeded the firing thresholds of the PTNs and were followed by spikes. In contrast, for slow PTNs, stimulation of the BC produced longer-latency, smaller EPSPs that were insufficient to take the cells to firing threshold and generate spikes. PTNs

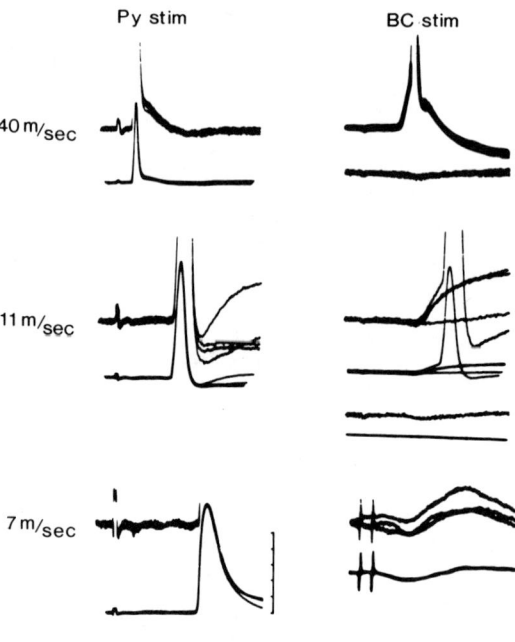

Figure 7.18. *EPSPs in large and small PTNs.* Left column shows antidromic responses to pyramidal tract stimulation and right column shows cerebellar-evoked EPSPs in PTNs with different axonal conduction velocities. The record at top right shows EPSPs in a relatively fast PTN (conduction velocity of 40 meter/sec) and the middle and lower records show the EPSPs in PTNs with slower axonal conduction velocities. Double-shock stimuli at lower right show temporal facilitation in the EPSP of the small PTN. In faster PTNs, stimulation of BC produces large EPSPs with short latencies, and these EPSPs commonly exceed firing threshold and are followed by spikes. In contrast, slower PTNs have EPSPs that are disynaptic and of smaller amplitude (middle and lower right). The lower traces in each record of the right column are field potentials recorded just outside the cells. Time calibration is 1 msec for all records except lower right, where calibration indicates 2 msec intervals. (From Shinoda, unpublished)

with intermediate conduction velocities had fairly large EPSPs with intermediate latencies. Figure 7.19 shows the relationship between the conduction velocity of PTNs and the latency of the EPSPs evoked by BC stimulation.

While well suited to provide information regarding different strengths of cerebellar signals (via VL) to PTNs of different sizes, BC stimulation cannot provide information on differences in the cortical effects of inputs from DEN as contrasted to IP. To obtain such information one must adopt a procedure analogous to that employed for contrasting the effects of DEN and IP on VL neurons. This requires careful placement of stimulating electrodes in DEN and IP to determine that stimulating current is not spreading

CEREBELLAR PATHWAYS TO FRONTAL AGRANULAR CORTEX 113

from one nucleus to the other. Even with perfect placement of stimulating electrodes in IP, however, it is difficult (as noted above) to rule out involvement of dentate fibers passing through the interpositus nucleus. Assuming selective stimulation of DEN and IP, however, it would be possible to record both field potentials and single-unit responses in MI, and to make inferences as to which individual MI PTNs receive convergent inputs from the two cerebellar nuclei.

Figure 7.20 provides a schematic representation of an experimental procedure to demonstrate these phenomena. Three sorts of recording electrodes are used:

1. A large electrode on the cortical surface to detect electrically evoked field potentials.
2. An intracellular electrode in a PTN to record EPSPs.
3. An extracellular electrode near a PTN to record PTN action potentials.

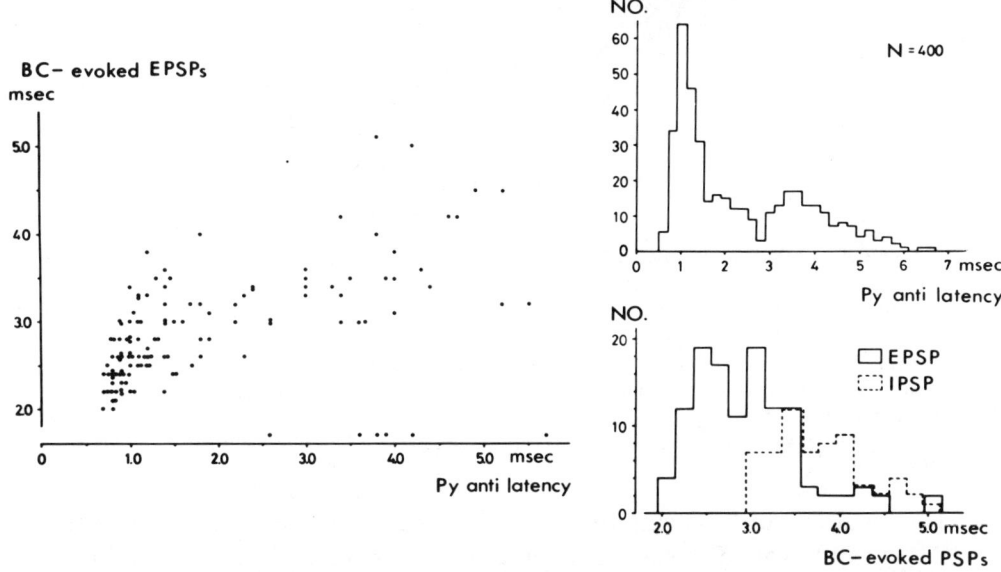

Figure 7.19. *Relation between antidromic spike latencies and cerebellar-evoked EPSPs in pyramidal tract neurons in the motor cortex.* Left: antidromic spike latencies from the medullary pyramid are plotted on the abscissa and latencies of EPSPs evoked from the brachium conjunctivum are plotted on the ordinate. The diagram indicates that fast PTNs receive short latency EPSPs from the brachium and slow PTNs receive longer latency EPSPs. Upper right: latency histogram of antidromic spikes in 400 PTNs evoked from the medullary pyramid. Lower right: latency histogram of EPSPs and IPSPs evoked from the contralateral brachium conjunctivum in PTNs. The shortest EPSP latency was 2 msec and the shortest IPSP latency was 3 msec. EPSPs with latencies of less than 3 msec were considered to be disynaptic from the brachium conjunctivum. (From Shinoda, unpublished)

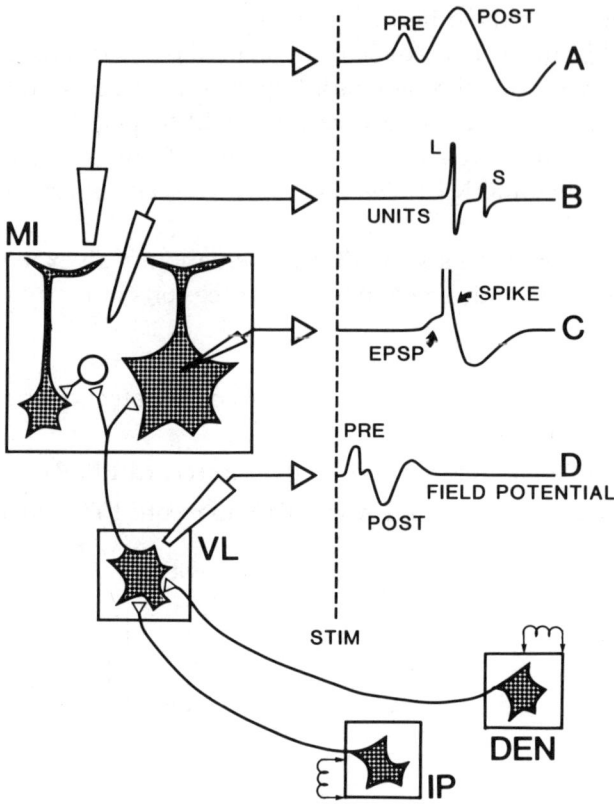

Figure 7.20. *EPSPs, single units, and field potentials.* A: A large electrode on the surface of the primary motor cortex (MI) picks up presynaptic (PRE) and postsynaptic (POST) field potentials evoked by stimulation of IP or DEN. B: An extracellular microelectrode positioned near two PTNs picks up spikes evoked by DEN or IP stimulation, a larger spike (L) being evoked in the larger PTN via a monosynaptic excitation from VL and a smaller spike (S) being excited disynaptically in the smaller PTN. C: The EPSP and spike are from an intracellular microelectrode. D: Field potentials can also be recorded from VL, as shown in the pre- and postsynaptic responses picked up from a large electrode following electrical stimulation of DEN or IP.

The result of studies using this procedure was an estimate that 90% of the PTNs in the cat MI receive convergent input from DEN and IP (Shinoda et al., 1982).

INTERPRETING FIELD POTENTIAL VARIATION AND CONVERGENCE OF CEREBELLOCORTICAL PATHWAYS

With this background on the neuroanatomical and neurophysiological information at the intracellular and systems levels, it is now possible to

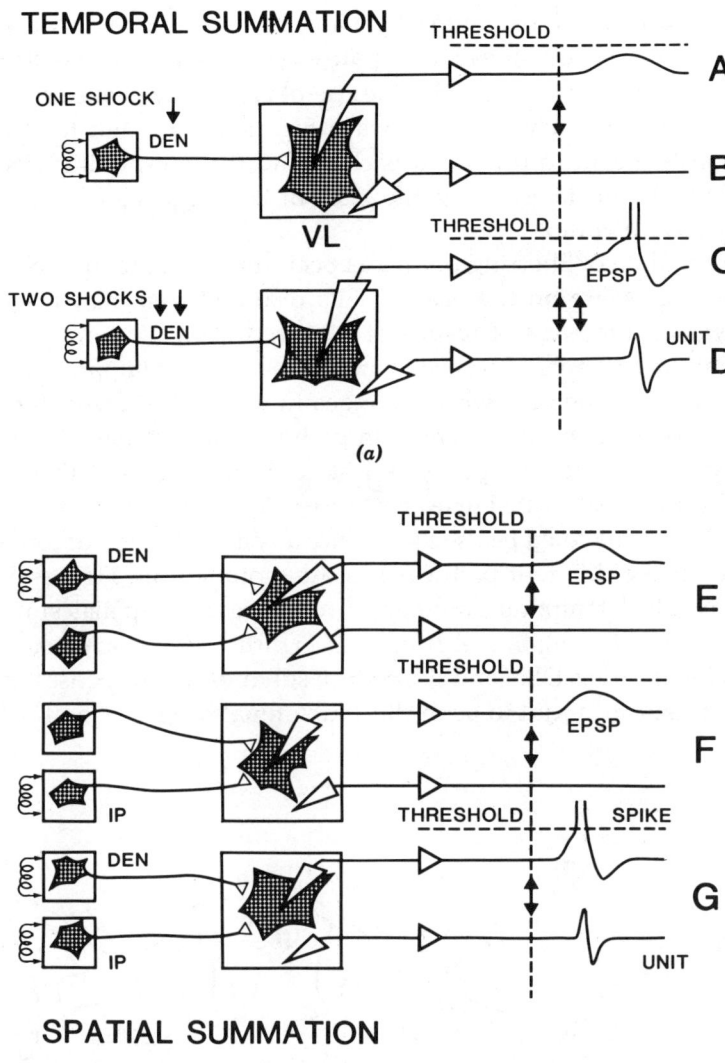

Figure 7.21. *Temporal summation and spatial summation.* Temporal summation has different manifestations depending on whether it is detected with intracellular or extracellular recordings. A: With intracellular recording, even a *subthreshold* input resulting from a single stimulus to DEN (arrow) will evoke an EPSP that will be detected by an intracellular electrode. B: An extracellular electrode placed near the same cell, however, will not pick up the consequences of this subthreshold input since the cell failed to fire. C: If two stimuli are delivered in close succession (arrows), the EPSP will reach threshold, generating a spike. D: As a result of the spike an extracellular electrode will also show the cell's activity. E, F, and G: Illustrate convergence of pathways from DEN and IP onto the same VL neuron. Stimulation of DEN alone (E) or IP alone (F) causes subthreshold EPSPs that fail to reach firing threshold for the neuron, but simultaneous stimulation of DEN and IP (G) will cause summation of EPSPs and as a result the VL neuron will fire.

115

consider how extracellular neurophysiological recording methods might be brought to bear on the question of pathways for set-related switching. In Chapters 8 and 9 we will consider the use of electrically evoked field potentials (FPs) in experiments on pathways for motor set, and the following discussion is meant to provide a background for interpreting FPs and to the use of FPs in determining the sites of convergence of DEN and IP pathways to MI cortex.

From the time of Sherrington, it has been traditional to infer changes in pathway excitability on the basis of FPs recorded from macroelectrodes, and it was on the basis of results of such recordings (often from ventral roots) that the concepts of occlusion and of spatial and temporal summation evolved. In Sherrington's work, changes in the magnitude of the ventral root FP were used to infer changes in pathway effectiveness; he delivered a variety of conditioning peripheral stimuli and followed these by test electrical stimuli that elicited reflex responses. A change in the reflex response to the test stimulus could be used to deduce an effect of the prior conditioning stimulus on the different pathways to the motoneuron. Figures 7.21 and 7.22 schematize temporal summation, convergence, spatial summation, occlusion, and the subliminal fringe. In Figure 7.23 the schematic sets of neurons identified as DEN and IP produce subthreshold responses; therefore the target neuron is said to be in the "subliminal fringe" of both DEN and IP.

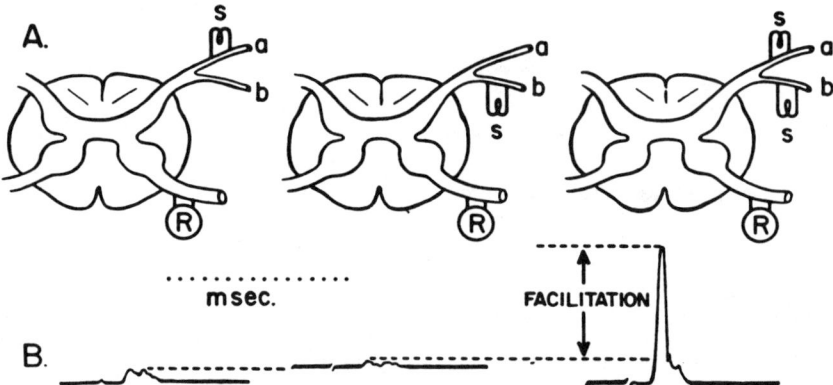

Figure 7.22. *Spatial facilitation demonstrated with field potential recording.* Recordings from a ventral root (at R) give a small field potential when fraction (a) of the dorsal route is stimulated (lower left) and another small field potential when fraction (b) is stimulated (lower middle). Since both (a) and (b) converge on the same motoneurons, simultaneous stimulation of (a) and (b) will produce marked spatial facilitation, that is, the field potential evoked by simultaneous stimulation of (a) and (b) is much larger (lower right) than the sum of the two field potentials evoked by individual stimulation of (a) or (b). (From Ruch and Fulton, 1960)

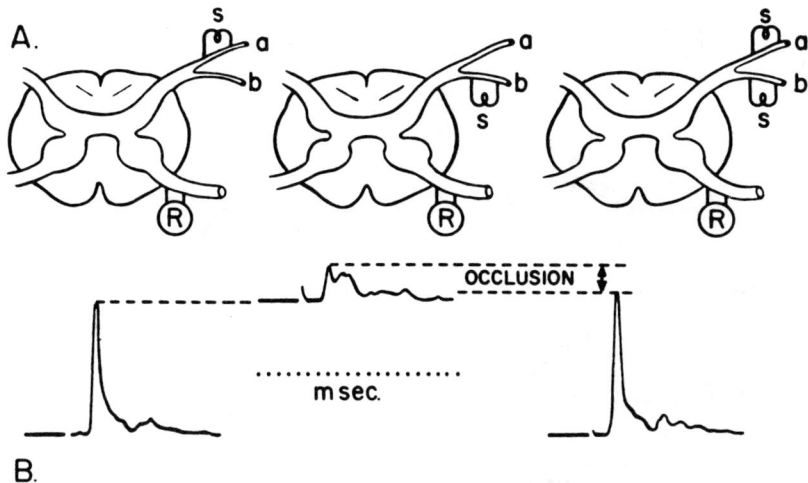

Figure 7.23. *Occlusion resulting from convergence of two input pathways on the same output elements.* If two input pathways to the same motoneurons are strongly stimulated, then some motoneurons will discharge in response to either pathway alone, and the response to simultaneous stimulation of the two pathways will be less than the sum of the individual responses to stimulation of each pathway alone. Conventions as in Figure 7.22. (From Ruch and Fulton, 1960)

The use of surface electrodes to pick up FPs has provided information as to *convergence* and *spatial summation* of DEN and IP signals in VL. As shown in Figure 7.24, both the small, positive deflection corresponding to *thalamic input* and the subsequent field potential in the cortex that are evoked by simultaneous stimulation of DEN and IP are greater than the algebraic sum of the potentials evoked by DEN and IP stimuli delivered separately. The effect on the earliest phase of the potential points to the *convergence* of IP and DEN signals at the level of VL neurons. The subsequent effects could indicate convergence at the cortical or thalamic level. Thus, a weak IP or DEN stimulus will excite a few VL neurons, leaving many others with subthreshold inputs. When these same weak stimuli are delivered simultaneously, however, many of those VL neurons that receive subthreshold inputs from separate stimuli within DEN or IP are driven over threshold. It has already been indicated (Figures 7.14, 7.15, and 7.16) that some inputs from DEN and IP converge on VL neurons, and the summation of FPs shown in Figure 7.24 can be predicted on the basis of this convergence. Summation of the sort shown in Figure 7.24 occurs with relatively weak stimuli. Indeed, for sufficiently weak stimuli to DEN and IP, there might be no MI field potential (FP) for separate stimuli, and yet a large FP might nonetheless appear when the two stimuli were delivered simultaneously. The sorts of conclusions reached on the basis of the FPs illustrated in Figure

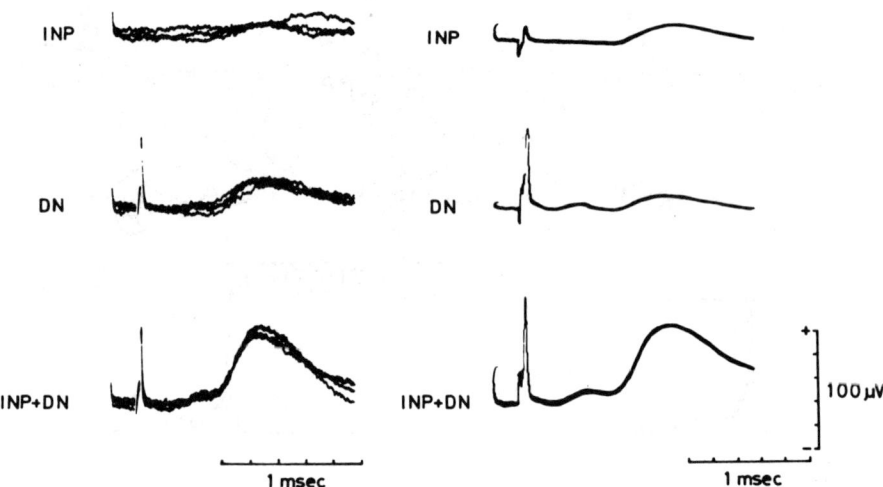

Figure 7.24. *Cerebellar-evoked MI field potentials.* Stimulation of the posterior interpositus (INP) and dentate nuclei (DN) evoked field potentials consisting of early small positive waves followed by later larger positive potentials shown in single superimposed traces (left) and averaged responses (right). The early positive waves reflect the action potentials of thalamocortical axons in the cortex and the later large positive waves mainly indicate synaptic currents produced by thalamic volleys. Note that the small positive deflection (indicating presynaptic thalamocortical inputs) when INP and DN are stimulated simultaneously (lower right) is greater than the sum of the positive deflections for INP and DN stimulated separately. This finding is consistent with convergence of DN and INP onto VL neurons in the thalamus. (From Shinoda, unpublished)

7.24 can also be arrived at via intracellular recording of EPSPs, a correspondence that is to be expected if we assume that the FP amplitude is based on the algebraic summation of the extracellular currents associated with EPSPs in many cortical neurons. Spatial summation (with weak stimuli) and occlusion (with strong stimuli) are shown for EPSPs in Figure 7.25, where combined strong stimulation of DEN and IP may evoke an EPSP whose amplitude is less than the algebraic summation of the EPSPs evoked by separate stimuli within the two nuclei. Occlusion results when the same VL neurons are caused to generate spikes either by separate stimulation of DEN or of IP. Comparable occlusive processes could also occur at the level of MI interneurons. It is to be noted that either spatial summation or occlusion points to the occurence of some convergence in VL and/or MI between terminals of neurons activated by stimuli at the two sites.

Figure 7.26 shows spatial summation (with weak DEN and IP stimuli) as reflected in the EPSPs of a single MI PTN. It is of particular importance for our consideration in the next chapter of the application of FP analyses to experiments on set-related behavior that, at least in certain cases, the conclusions based on FP analyses correspond to those based on intracellular EPSP recordings in antidromically identified cortical neurons. For our present discussion, however, it is sufficient to note our conclusion that convergence of DEN and IP inputs to MI could occur at both the thalamic and cortical levels in the cat, and may also occur in the monkey.

Cerebello-thalamocortical convergence may also be demonstrated by recording from the corticofugal fibers in the pyramidal tract. Amassian and Weiner (1966) used recordings of medullary pyramidal tract FPs and unit discharges as a measure of cortical outputs elicited by inputs that reached MI after VL output axons were electrically stimulated. It is common practice to implant permanent stimulating electrodes in the medullary pyramid for

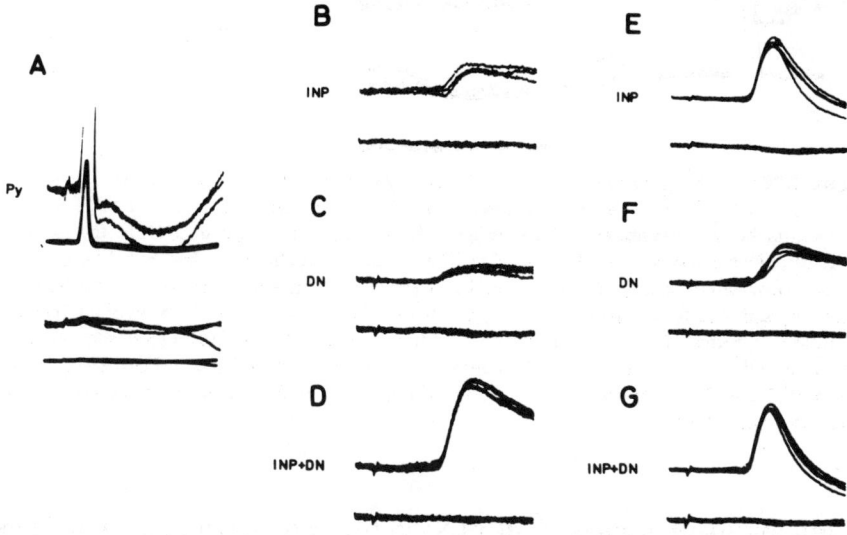

Figure 7.25. *Spatial facilitation and occlusion in intracellularly recorded VL responses to dentate and interpositus.* Both spatial summation and occlusion are based on convergence of inputs from two different sources, and—depending upon stimulus strength—a given site may show spatial facilitation or occlusion. The central column of traces (B, C, and D) illustrate spatial facilitation, whereas the column of traces at the right (E, F, and G) show occlusion. A: Antidromic identification of relatively fast pyramidal tract neuron. B and C: Disynaptic EPSPs evoked by weak stimulation of the interpositus (B) and the dentate nucleus (C). D: Stimuli were applied simultaneously to both nuclei. The amplitude of EPSP in D is much larger than the algebraic sum of individual EPSPs in B and C. This supports the view that some convergence of the inputs from both nuclei occurs at the level of VL. E and F: Disynaptic EPSPs evoked following stronger interpositus (E) and dentate (F) stimulation. G: Simultaneous stimulation of both nuclei produced occlusion, as shown by the fact that the EPSP amplitude of G is smaller than the algebraic summation of EPSPs in E and F. This finding confirms convergent inputs to single VL neurons from both interpositus and dentate nucleus as shown by EPSPs from VL neurons. The lower traces in B–G show the negligible field potentials just outside the cell, and validate the intracellular EPSPs. (From Shinoda, unpublished.)

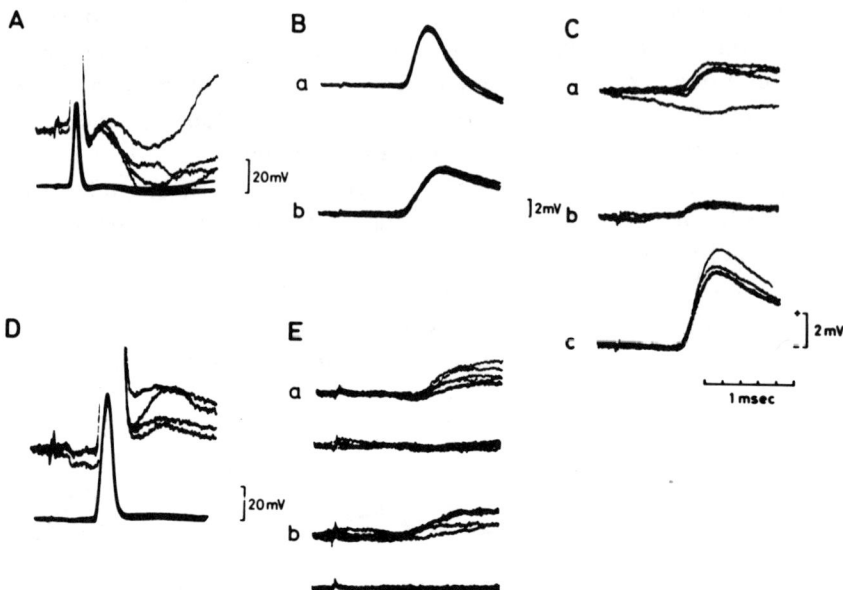

Figure 7.26. *Convergence of dentate and interpositus outputs to a fast (A–C) and a slow (D–E) pyramidal tract neuron in the motor cortex.* A: Antidromic activation of a fast PTN recorded intracellularly. B: Disynaptic EPSPs evoked by stimulating IP (a) and DEN (b) (stimulus intensity, 500 microamperes). C: Spatial facilitation between the IP-evoked EPSP (40 microamperes) (shown in a) and the DEN-evoked EPSP (300 microamperes) (shown in b). Both stimuli were applied simultaneously in c. D: Antidromic spike of a slow PTN evoked from the medullary pyramid. E: Polysynaptic EPSPs induced (a) by IP stimulation (300 microamperes) and (b) by DEN stimulation (300 microamperes). The lower traces in a and b are extracellular field potentials. The amplitude calibration in C applies to all high gain traces except B. (From Shinoda et al., 1982)

antidromic identification of MI PTNs in behaving monkeys, but little use has been made of these chronically implanted macroelectrodes for recording FPs associated with electrically evoked discharges descending from MI. Much of the descriptive work on such electrically evoked FPs in the pyramidal tract was summarized by Patton and Amassian (1960), who showed that the FP evoked by electrical stimulation of MI contained a direct, nonsynaptic component resulting from activation of descending pyramidal tract axons, as well as a series of indirect components due to activation of intracortical elements that, in turn, led to synaptic activation of PTNs. Stimuli applied to white matter after extirpation of the cerebral cortex continued to evoke the direct wave, but not the indirect waves.

Of particular relevance to the issues of set-related gating are the results of Amassian and Weiner (1966, p. 261) on factors affecting the amplitude of the pyramidal tract FP evoked by VL stimulation. It was noted that:

> The marked variability in both latency and amplitude of the relayed PT [pyramidal tract] response under conditions of prominent "spontaneous" discharge by PT neurons suggests the possibility that such variability is related to antecedent activity in PT neurons and associated interneurons.

Those investigators confirmed the findings of Asanuma (1959) that the relayed pyramidal tract response to VL shock could be enhanced by prior stimulation of forelimb afferent nerves, and stated that (Amassian and Weiner, 1966, pp. 262–263):

> The conditioning effect of a peripheral stimulus might be exerted at the thalamic level by lowering the threshold of directly activated corticipetal axons, or at the cortical level by convergence either on pyramidal tract neurons or on intercalated neurons. However, facilitation of the first relayed response occurred when the test stimulus was delivered anterior to and lateral to VL and VA. Any depolarizing effect of the peripheral stimulus would presumably be greatly diminished at such distance from the somata of VL and VA neurons. We conclude that facilitation of the first relayed response by peripheral stimulation is exerted, in part at least, at the cortical level.

These older studies point to the possibility that the pyramidal tract FP evoked by electrical stimuli that activate VL pathways into MI might reveal shifts in the effectiveness of those pathways that converge on MI neurons as a function of changed motor set (see Chapters 8 and 9).

CEREBELLO-THALAMOCORTICAL ELECTROPHYSIOLOGY: SYNTHESIS

Of the various data we have considered, perhaps the most intriguing is the relative weakness of the excitation of small PTNs from the cerebellum. For a given shock intensity, larger PTNs have larger EPSPs (and are more likely to generate spikes) than are smaller PTNs. This comes as a surprise, because we know from studies in behaving monkeys that smaller PTNs share certain properties with smaller motoneurons, and thus tend to follow a "size principle" (see Henneman, 1974). For example, smaller PTNs are tonically active at rest, whereas larger PTNs are silent (Evarts, 1965). Furthermore, smaller PTNs are recruited earlier than large PTNs (Evarts et al., 1983) and are more intensely driven by afferent input (Evarts and Fromm, 1977). Thus it would appear that although the smaller MI PTNs are "turned off" in the acute cat preparations that have been used for the electrophysiological studies, they are quite the opposite in the behaving monkey. But this very fact points to a possible mechanism for switching in MI, and that the smaller PTNs are separated from thalamic inputs by cortical interneurons points to the possibility of set-related interneuronal gating in MI. In experiments on set-dependent gating, one of the first questions one would ask would pertain to set-related changes in PTN responsiveness to VL inputs as a function of PTN size (see Chapter 10).

This discussion of small PTNs with their apparent paucity of monosynaptic inputs from VL reminds us that neurophysiologists have a special fondness for monosynaptic inputs because such inputs are amenable to experimental analysis. Students of neurophysiology sometimes get the mistaken impression that because monosynaptic inputs are more accessible to investigation they are more important in the functioning of the nervous system. In the case of the monosynaptic projection from VL to large MI PTNs, for example, one could mistakenly assume that this input was the most important pathway whereby VL exerted its influence on large PTNs, and that polysynaptic pathways were unimportant. But the *presence* of monosynaptic inputs from VL to large MI PTNs does not mean that those PTNs lack polysynaptic inputs from VL. There is also a monosynaptic connection with motoneurons for PTN output to spinal cord, but there is no reason to assume that this connection is more important than are the numerous polysynaptic connections via the spinal cord interneuronal networks. The point to be made here is that, although monosynaptic connections are extremely valuable for localizing the site of a dynamic process in neurobiological studies, such synapses should not be thought of as more powerful or more significant than polysynaptic links. From the standpoint of set-related switching, it would seem that the links via interneurons could actually be more important.

PALLIDO-THALAMOCORTICAL AND NIGROTHALAMOCORTICAL ELECTROPHYSIOLOGY

Many areas of neurophysiology are virtually overwhelmed with a flood of new data, but an outstanding exception is in the electrophysiology of the systems that link globus pallidus with thalamus and thence with cerebral cortex. Indeed, much of the work on the electrophysiology of pallidothalamic connections was done 15 or 20 years ago, and then, although many questions remained unanswered, the pace of research slackened. A review by Kitai (1981) provides a treatment of pallido-thalamocortical connections and raises two crucial questions on pallidothalamic electrophysiology: Is the synaptic action of pallidal terminals in thalamus excitatory or inhibitory, and do pallidal terminals converge on the same thalamic neurons that receive inputs from the cerebellum? The initial papers on these topics (Desiraju and Purpura, 1969; Dormont and Ohye, 1971; Malliani and Purpura, 1967) reported that stimulation of pallidal output systems evoked monosynaptic excitation of some VL and VA neurons, and that a few individual neurons received convergent inputs from cerebellum and pallidum. However, in a reexamination of these issues (Uno and Yoshida, 1975; Uno et al., 1978) it was found that pallidal outputs evoked monosynaptic *inhibition* and that there was no convergence from cerebellum and pallidum onto the same neurons. In concluding their report Uno and Yoshida (1975, p. 380) stated that:

> . . . our result shows physiologically that the distribution of the pallidothalamic fibers within the thalamus is in the more ventrorostral part in the VL-VA complex as compared to the area where the cerebello-thalamic pathway terminates. Thus the pallidal fiber exerts inhibitory influences on the cell group which is rather different from thalamic relay neurons of the cerebello-thalamo-cortical pathway, and it seems likely that both cerebellar and pallidal influences do not converge directly on a single thalamic neuron, or if any, converge on a very few cells.

These results are supported by anatomical studies (Hendry et al., 1979; Asanuma et al., 1983). A recent study by Deniau et al. (1982) has extended the findings of Uno and Yoshida to the nigrothalamic pathway. In this work, electrical stimulation of the pars reticulata of the substantia nigra evoked monosynaptic IPSPs in thalamic neurons.

In summarizing the overall signal flow through basal ganglia (Figure 7.27), Kitai (1981, p. 1012) speculates that:

> The striatal projection neurons exert inhibitory action on neurons of the globus pallidus and the substantia nigra (pars reticulata). The projection cells of the substantia nigra (pars reticulata) and the globus pallidus (medial segment) are also inhibitory to their target thalamic neurons (VA, VL and VM). The thalamic neurons, in turn, excite those cortical neurons with axons projecting to brain stem and spinal cord motor center. . . .
>
> Thus, the major outflow of signals from the striatum to the thalamus is channeled mainly through two structures, pars reticulata of the substantia nigra and the medial segment of the globus pallidus. Since the synaptic actions of the striatal, pallidal, and nigral (pars reticulata) neurons are all inhibitory, any signals exciting the striatal neurons (e.g. cortical inputs) are expressed in the thalamus as excitation via the mechanism of disinhibition. That is, an excitation of the striatum results in release of inhibition on the thalamocortical system.

In Kitai's admittedly speculative formulation, excitatory inputs from cerebral cortex to the striatum (e.g., from prefrontal cortex to caudate nucleus or, for our purpose, from the somatic sensorimotor cortex to the putamen)

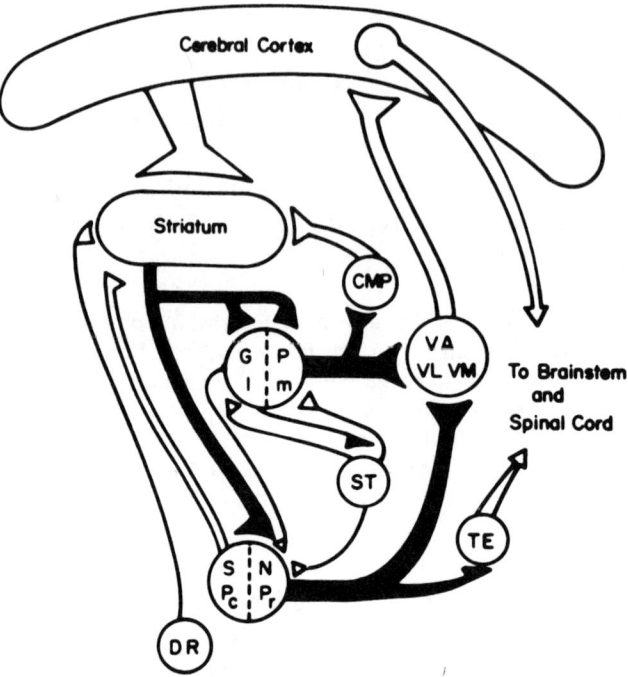

Figure 7.27. *Summary of neuronal circuitry in basal ganglia and related structures.* Open terminals are excitatory connections and filled terminals are inhibitory connections. CMP, centre median-parafascicular complex; DR, dorsal raphe nucleus; GP, globus pallidus; l, lateral; m, medial, P_c, pars compacta; P_r, pars reticulata; SN, substantia nigra; ST, subthalamus; TE, superior colliculus (tectum); VA, ventroanterior nucleus of thalamus; VL, ventrolateral nucleus of thalamus; VM, ventromedial nucleus of thalamus. (From Kitai, 1981)

would have the net effect of *reducing* impulse frequencies of pallidal output neurons and this, in turn, would result in *disinhibition* of thalamocortical neurons relaying pallidal output signals to cerebral cortex.

In our discussions of the cerebello-thalamocortical circuit we could provide a systematic review of the electrophysiology of each step in the pathway and, indeed, could go further back along the path to consider the synaptology of inputs to IP and DEN from cerebellar cortex and brainstem, the inputs from inferior olive, pontine nuclei, and ascending tracts to cerebellar cortex. But when we turn to the physiology of thalamic neurons that receive pallidal inputs and ask how these neurons act on the cortex, we have a scarcity of information. There are simply no studies comparable to those showing how DEN and IP converge on VL neurons, which in turn produce both monosynaptic and polysynaptic EPSPs in cortical neurons. This lack of electrophysiological information is especially unfortunate when one seeks to study pathways underlying set-related behavior, because the supplementary motor cortex, which receives pallidal signals via thalamus, contains

"NONSPECIFIC" THALAMOCORTICAL PROJECTIONS

many set-related neurons and projects to MI. In Chapter 8 we will consider use of electrical stimuli to test for set-related changes within the cerebello-thalamocortical pathway, and it would be desirable to outline an analogous approach for the pallido-thalamocortical pathway. It would seem clear that a high priority in work on mechanisms of higher brain function should be acquisition of basic electrophysiological information on the pallido-thalamocortical projection systems in the primate.

"NONSPECIFIC" THALAMOCORTICAL PROJECTIONS

In addition to their projections to the VL complex, the cerebellar nuclei and globus pallidus project to other thalamic nuclei that, in turn, project to cerebral cortex. These nuclei (often referred to as "nonspecific") provide a second major route whereby cerebellar and pallidal outputs can influence cerebral cortex. The nonspecific thalamocortical systems are of special interest in the context of experiments on set in view of the finding by Purpura and Housepian (1961) that, in the cat, suitably timed low-frequency stimulation of certain putatively *nonspecific* thalamic nuclei markedly attenuates *specific* MI responses evoked by stimulation of pathways below VL. These inhibitory

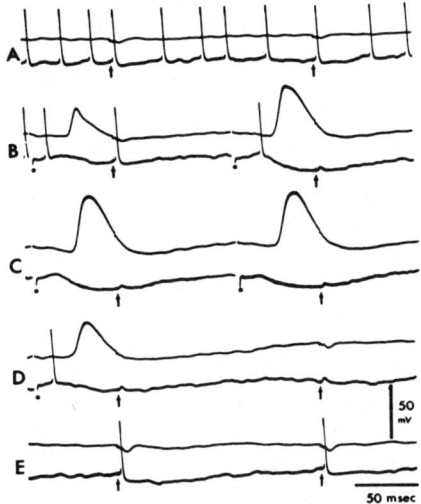

Figure 7.28. *Thalamothalamic interactions.* In addition to receiving inputs from the cerebellar nuclei via the brachium conjunctivum (BC), VL neurons receive inputs from other parts of the thalamus, for example, the reticular complex. Some of these are inhibitory, as illustrated in this set of records showing inhibition in a 'L cell. A: the cell exhibits considerable spontaneous activity. B, C, D, and E: activity during BC stimulation (arrows) combined with medial thalamic stimulation (dots). Note the decrease in activity which lasts nearly 200 msec after last stimulus. Nonspecific inhibitory effects such as these may play a role in gating inputs to motor cortex. (From Purpura et al., 1965)

effects of the nonspecific thalamocortical system on MI responses relayed by VL provide a potential substrate for set-related gating in VL. Inhibitory postsynaptic potentials (IPSPs) that might underlie thalamically mediated gating of cerebellocortical signals are shown in Figure 7.28.

Experiments on nonspecific systems have failed to provide insights that were hoped for in the 1950s and 1960s, when study of the "reticular activating system" and the "nonspecific thalamocortical projection systems" were in vogue. Hobson (1980, p. 7) has commented on the virtual disappearance of electrophysiological research on the reticular formation, and his remarks apply equally well to electrophysiological research on the nonspecific thalamocortical systems:

> In the 20 years since the last major publication on the reticular formation, the interest of the neurobiological community has waned almost to the point of selective inattention. The reasons are clear. The dramatic effects of stimulation and lesions clearly indicated important ascending and descending influences arising from and passing through the reticular formation; it was obvious from the outset that a specification of those influences was crucial to our understanding of the mechanisms of sleep and waking. The lesion, stimulation, and electroencephalogram (EEG) recording methods, however, were incapable of specifying the functions of the reticular formation in cellular neurophysiological terms.
>
> Thus the black arrows shown shooting up and down the neuraxis raised unanswerable questions which soon failed to sustain scientific attention. One segment of the neurobiological community, the sensorimotor physiologists, asked: (a) Where did the arrows actually begin and end? (b) How many different kinds of arrows were there? (c) What did the pluses and minuses placed at the tips of the arrows mean? (d) What chemicals coated the heads of the arrows?
>
> Another segment, the behavioral state physiologists, wondered: (a) Did the arrows always point in the same direction? (b) Were the arrows shot with the same strength at all times? (c) Did the arrows always have the same effect when they hit their targets? (d) What determined whether or not the arrows were shot?
>
> The failure to answer these questions not only led to disinterest but also promoted the misleading ideas that the reticular formation was either undifferentiated or too complex to be studied, or both. Ironically, the period of declining interest in the reticular formation was marked by the development of the very techniques and concepts needed to answer many of the questions raised above.

We might ask, in parallel with Hobson's remarks, whether declining interest in the classical, if somewhat "old-fashioned," techniques in the domain of neurophysiology has occurred just when they might make a most significant contribution.

Chapter 8

Experiments to Identify Pathways Involved in Set-Related Switching

Earlier chapters in this monograph have reviewed set-related cortical activity and the anatomy and electrophysiology of the pathways that may mediate this activity. This chapter considers the design of experiments in which behavioral and electrophysiological approaches are combined to study the dynamic features of pathways that underlie set. Some of the ideas that are presented emerged from the first of The Neurosciences Institute conferences mentioned in the Preface.

The work of Tsukahara, showing that classical conditioning and nerve cross-union modify the effectiveness of the corticorubral pathway, has provided paradigms in which field potentials and electrically evoked single-unit discharges have been shown to change in parallel with intracellularly recorded EPSPs.

CORRELATIONS BETWEEN EPSPs AND FIELD POTENTIALS IN RED NUCLEUS

Combined intracellular and field potential (FP) recordings in red nucleus were obtained by Tsukahara et al. (1982) in cats before and after forelimb cross-union, that is, switching the nerves that control two muscles in the upper limb. In these experiments, electrically evoked postsynaptic FPs recorded with macroelectrodes revealed altered effectiveness of synaptic inputs to red nucleus after the cross-union. In parallel studies that used intracellular recordings, it was observed that cross-union of upper limb nerves was followed by a change in rise time of EPSPs evoked in rubrospinal neurons by electrical stimulation of cerebral motor cortex or cerebral peduncle, but without any alteration of EPSPs evoked by interpositus nucleus stimulation. After the nerve cross-union in the upper limb, cerebral peduncle stimulation evoked EPSPs with faster rise times in red nucleus cells innervating the cervical cord, but not in those innervating the lumbosacral cord.

These results are relevant to the topic of this chapter because they demonstrate an effect of cross-union in terms of both extracellularly recorded FPs and intracellularly recorded EPSPs. The effect was a change in the time-course of facilitation of the FP by a conditioning (i.e., preceding) shock to the cerebral peduncle. Interpositus (IP) stimulation evokes a prominent FP in red nucleus, and this FP has clearly distinguishable presynaptic and postsynaptic components, as shown in Figure 8.1. When a test stimulus to IP was preceded by a conditioning stimulus to the cerebral peduncle, the postsynaptic component of the red nucleus FP evoked by the IP shock was facilitated, whereas the presynaptic component of the FP was unchanged. The time-course of the facilitation of the postsynaptic FP component was measured, and it was found to be in good agreement with the time-course

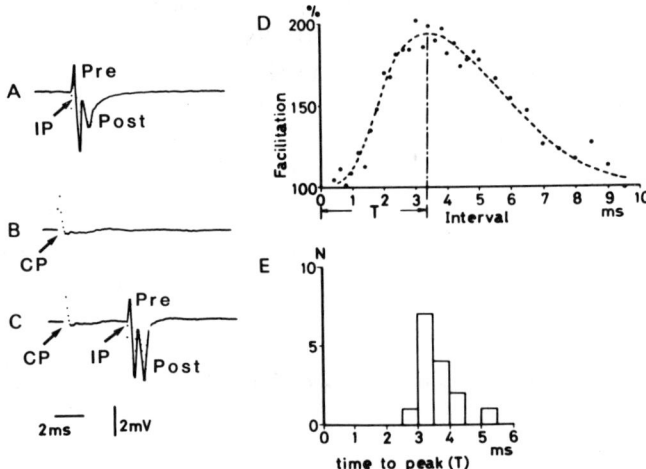

Figure 8.1. *Red nucleus field potentials evoked by interpositus (IP) stimulation.* A: Electrical stimulation of IP evokes a field potential in which there is a presynaptic (Pre) component followed by a postsynaptic component (Post). B: The brief initial deflection is a shock artifact occurring with stimulation of the cerebral peduncle (CP), but the excitation resulting from this CP shock is insufficient to generate a clearly identifiable field potential. C: However, if such an apparently ineffective CP shock precedes an interpositus stimulus, then a facilitatory interaction is observed. The magnitude of this facilitation of the postsynaptic field potential varies depending upon the interval between the conditioning CP stimulus and the subsequent IP stimulus, and the interval at which facilitation is maximal is referred to as the "time-to-peak facilitation." D: Results of an experiment in which facilitation reaches a maximum when the two stimuli are separated by about 3.3 msec. E: Results of the same experiment in a number of different cats with different electrode placements indicates that for these different experiments the time-to-peak facilitation was rarely less than 3 msec and was commonly between 3 and 4 msec. (From Tsukahara et al., 1982)

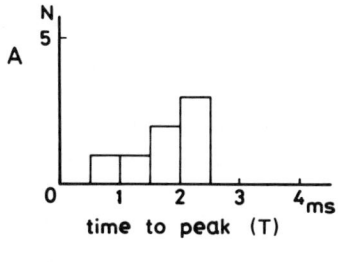
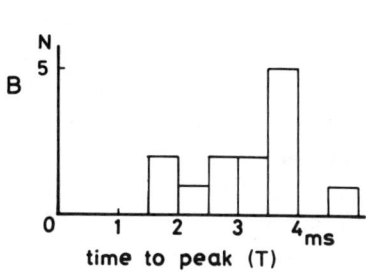

Figure 8.2. *Effects of cross-innervation on field potential facilitation.* Data derived as shown in Figure 8.1E were obtained from different parts of the red nucleus for cross-innervated cats. Histograms of "time-to-peak facilitation" in the "cervical" (A) and "lumbar" (B) regions of the red nucleus illustrate that for the cervical region the time-to-peak facilitation was considerably earlier than usual (commonly less than 3 msec), whereas in the lumbar portion of the red nucleus the time-to-peak facilitation was usually more than 3 msec, about the same as in normal cats. (From Tsukahara et al., 1982)

of intracellularly recorded red nucleus EPSPs evoked by stimuli to the cerebral peduncle: both the facilitation of the FP and the EPSP evoked by a cerebral peduncle shock reached a peak at about 3 msec. It therefore seemed reasonable to suppose that the time-course of facilitation of the postsynaptic FP depended on the time-course of the EPSPs evoked by the conditioning shock to the cerebral peduncle, and Tsukahara reasoned that this relationship might allow changes in the time-course of FP facilitation to provide indirect inferences as to changes in the time-course of EPSPs. Effects of cross-innervation on the time-to-peak facilitation were therefore studied, with the conditioning stimulus being applied to the cerebral peduncle and the test stimulus applied to the interpositus nucleus before and after nerve cross-union in the forelimb. FPs in regions of red nucleus that project to cervical zones of spinal cord (and therefore control the cross-innervated upper limb) were compared to FPs from parts of red nucleus projecting to the lumbar segments of the spinal cord, and a difference was found between times-to-peak facilitation in these areas after the cross-innervation. Figure 8.2 shows that the time-to-peak facilitation was enhanced for FPs in the rubrocervical parts of the red nucleus after cross-innervation, but not in the rubrolumbar parts. In commenting on these findings, Tsukahara et al. noted that although the time-course of the corticorubral facilitation of the red nucleus FP evoked by interpositus stimulation provides only an indirect method for evaluating the time-course of the EPSPs from corticorubral inputs, it has the advantage of reflecting the activity of a population of intact red nucleus cells whose behavior is not distorted by microelectrode impalement. A more rapid rise time of facilitation of the interpositus-evoked FPs thus provided supportive evidence for the previously observed changes in time-course of EPSPs evoked by stimulation of the cerebral peduncle after cross-innervation. This strengthened the idea that input-output relations of rubrocervical cells exhibit adaptive plasticity after cross-innervation of the upper limb nerves.

In considering Tsukahara's results in the context of experiments to identify pathways mediating short-term, set-dependent changes of input-output relations, it is important to note that he succeeded in obtaining significant diffcrences in spite of being unable to use a given recording site as its own control: that is, recordings before and after cross-union were either from different cats or from different loci in the same cat.

Rapidly occuring changes that reflect changes in set would be easier to study, since the speed with which these changes take place would allow potentials to be recorded from the same electrode both before and after the change of set. This would mean that, rather than relying entirely on

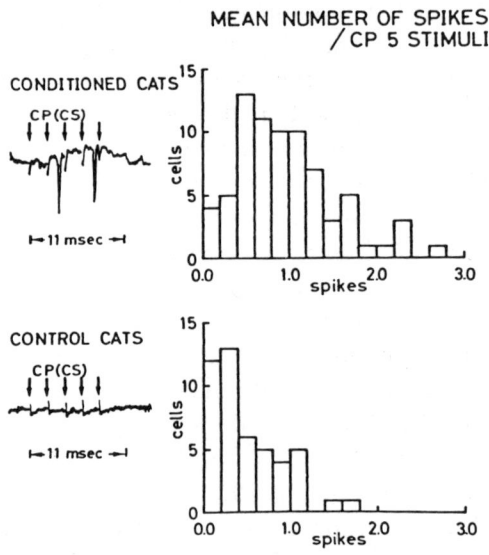

Figure 8.3. *Modification of red nucleus unit activity by classical conditioning*. As shown in Figure 8.1, stimulation of the cerebral peduncle (CP) evokes little if any field potential in the red nucleus. A comparable finding can be observed at the level of extracellular impulse recording, that is, CP stimuli evoke no impulses in control animals (lower left). However, in animals that have undergone classical conditioning (see text), a comparable train of stimuli to CP evokes two single unit impulses (upper left). To establish statistically significant differences between the probability of cell discharge in conditioned and control cats, many cells were examined and, as shown in the right pair of histograms, the mean number of impulses per set of five stimuli to CP was greater after conditioning (upper right) than before conditioning (lower right). (From Oda et al., 1981)

time-course changes (as in Tsukahara's studies on plasticity after nerve cross-union), recordings of set-related changes might also detect significant differences in amplitude of the postsynaptic component of the response.

In another set of experiments, Tsukahara et al. (1981) have shown that extracellularly recorded spikes may provide a correlate of the EPSP changes that occur in association with classical conditioning mediated by the red nucleus in the cat. In this study, the conditioned stimulus was an electrical stimulus applied to the cerebral peduncle and the unconditioned stimulus was a stimulus to the skin of the forelimb that produced limb flexion. After the conditioned and unconditioned stimuli were paired in close temporal association, an initially ineffective conditioned stimulus (a cerebral peduncle stimulus) came to elicit flexion of the elbow. It was shown that extracellularly recorded red nucleus unit responses to cerebral peduncle inputs were enhanced as a result of conditioning (Figure 8.3).

The authors reasoned that, since the threshold for elicitation of elbow flexion and the strength of elbow flexion induced by stimulation of the interpositus nucleus of the cerebellum were identical in the experimental and control animals, the interposito-rubrospinal system could not be the site of the plastic change with conditioning, and that modifications of the corticorubral synapses must therefore have been responsible for the behavioral change.

IMPLICATIONS OF WORK ON STRUCTURALLY BASED, LONG-TERM ADAPTIVE PLASTICITY FOR WORK ON DYNAMICALLY BASED, SHORT-TERM SET

How can Tsukahara's methods and conceptual approaches be applied to work on motor set in behaving monkeys? In beginning to answer this question, it will be useful to recall Tsukahara's observations that coordination of the cross-innervated forelimb was present during voluntary goal-directed movement but *not* during more automatic movements (e.g., locomotion). He speculated that for volitional movements in which cortex was selectively involved, the sprouting of corticorubral neurons onto the proximal portion of red nucleus dendrites after cross-innervation might allow adaptive plasticity for movements driven by cerebral cortex, but that for the more stereotyped movements of locomotion (controlled mainly by cerebellar pathways), the unaltered interpositorubral pathways would control red nucleus output and there would be little evidence of an adaptive change.

In the speculations of Tsukahara et al., the mechanism of the adaptive plasticity was presumed to be structural (synaptic growth), and of course this mechanism is different from the evanescent dynamic processes that underlie changes of motor set. What differences in experimental design are called for in studies, such as Tsukahara's, seeking to locate *structural change* or long-term changes in synaptic strength underlying adaptive plasticity versus studies attempting to locate *dynamic changes* underlying motor set? One important difference is that a key element in Tsukahara's experiments was to compare the synaptic effects of two input pathways ending *monosynaptically* on the same rubrospinal neuron. A shift in the relative effectiveness or time course of the synaptic effects of these two monosynaptic pathways is not likely to be attributed to general changes in the excitability of the postsynaptic cell itself, since such a change would be expected to modify the response of the cell to *both* input pathways. Though Tsukahara could not record from the same single cell both before and after cross-innervation or classical conditioning, he could compare the responses of the same cell to electrical stimulation of two input pathways and show that the relationship between responses changed as a result of cross-innervation or conditioning.

A change of effectiveness of one monosynaptic pathway relative to another as observed by Tsukahara for adaptive plasticity, is *not* to be anticipated for set-related changes in the cerebral cortex. Such an alteration of monosynaptic input strength could be imagined if presynaptic conductance changes occurred, but axoaxonic synapses on axonal terminals are not common in the neocortex. In our opinion, other sorts of local circuits are more likely to underlie set-related changes involving alteration in the relative effectiveness in one of two pathways.

Figure 8.4 illustrates such a local circuit for gating of transmission via an interneuron that is itself impinged on by a tonically active, set-related, inhibitory interneuron, and contrasts this circuit with the one hypothesized by Tsukahara to underlie the long-term changes in synaptic efficacy he observed. In Chapter 6, we reviewed experimental data showing that set-related neurons could be divided into two classes: those with tonic set-related activity in the interval between an instruction stimulus (IS) and a trigger stimulus, and those in which the effects of set could be observed in transient activity. In Figure 8.4, it is assumed that the inhibitory interneuron becomes tonically active during set B and reduces the effect of one of two pathways to a PTN. Note that a PTN output change identical to the one pictured in Figure 8.4 could be achieved by making the interneuron excitatory rather than inhibitory, and by turning it on for set A and off for set B. Figure 8.5 shows the hypothetical case of a set-related inhibitory interneuron projecting to a PTN, but unlike that shown in Figure 8.4, the set-related changes will be nonselective, affecting responses to electrical stimulation of *both* of the pathways to the PTN.

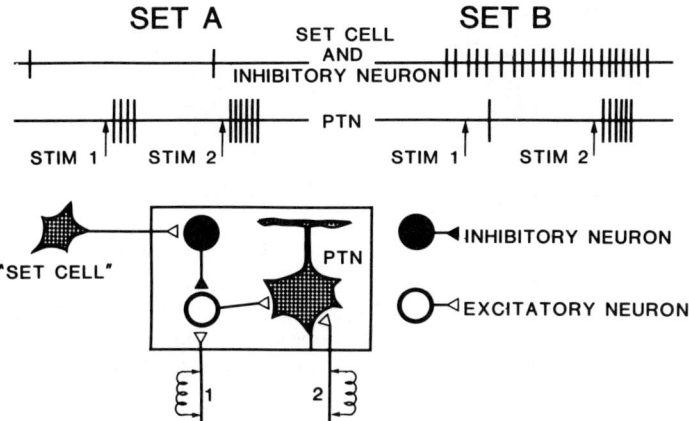

Figure 8.4. *Dynamic changes in pathway effectiveness.* In contrast to the change in pathway effectiveness illustrated in Figure 8.3 and by the work of Tsukahara, this figure illustrates a hypothetical local circuit that could change the effectiveness of a pathway depending on motor set. Pathway #1 is presumed to be effective during set A and ineffective during set B. Such a decrease in pathway effectiveness might result from tonic discharge of a "set cell" (indicated here as a single cell, but to be considered part of a circuit). The tonic discharge mediated by an inhibitory interneuron would diminish the response of the PTN to a stimulus applied to pathway #1 during set B. Note that the PTN, the output of which provides an index of the effectiveness of the two pathways, is also impinged on by a second pathway, referred to as pathway #2. Changes in activity of the inhibitory interneuron would not modify the effectiveness of pathway #2 in generating outputs from the PTN. This second pathway, presumed not to change as a function of set, is analogous to the pathway that Tsukahara investigated from interpositus to red nucleus. Thus, Tsukahara found a change in effectiveness of one pathway (from cerebral peduncle to red nucleus) but no change in the effectiveness of the pathway from interpositus to red nucleus.

Probably the major point in common between Tsukahara's experiments and those that might be carried out on motor set is the shift of control of an output neuron from one pathway to another. Tsukahara's experiments involved switching that gave relatively greater control of red nucleus to cerebral cortex. Even earlier, Tsukahara had considered shifts in the role of interpositus with changes of motor strategy (Allen and Tsukahara, 1974), proposing that the onset of a centrally programmed movement, as compared to a feedback-regulated movement, might involve a switching of control of MI output from feedback control, based on interpositus, to mainly preprogrammed control based on dentate. To generalize this idea, one may designate sites providing MI with feedback (FB) as FB_1, FB_2, ..., Fb_n. In addition to interpositus, these various FBs might include certain subdivisions of the primary somatosensory cortex in the postcentral gyrus, the second somatosensory area, and various thalamic nuclei (see Chapter 7). Hypothetical sites for centrally programmed (CP) inputs to MI might be designated as

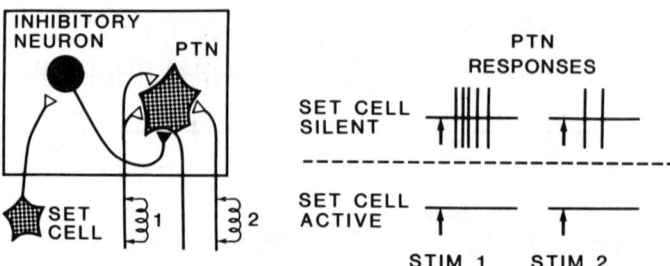

Figure 8.5. *Nonselective gating.* The set cell circuit and inhibitory interneuron projecting to the PTN are silent during one set and active during another set. Unlike the case of Figure 8.4, however, the set-related changes will be nonselective, affecting responses to electrical stimulation of *both* the pathways to the PTN.

$CP_1, CP_2, \ldots, CP_n$. In addition to dentate, these various CPs might include the supplementary motor cortex, premotor cortex, and certain thalamic sources. Let us now assume that a trigger stimulus (TS) that follows an instruction stimulus (IS) will cause the animal to switch from what we might call the feedback mode, important in maintaining stability, to the more centrally programmed mode essential for making a large, high-velocity limb movement. If we hypothesize that this involves a shift from FB to CP control, then field potentials and unit activity evoked in MI by electrical stimulation of one of the sites that relays feedback to MI might become smaller after the TS, whereas field potentials and unit activity evoked by electrical stimulation of one of the sites sending central programs to MI might become larger. This *hypothesis* of enhanced MI responses evoked from a site sending central programs to MI may be seen to parallel Tsukahara's *observation* of enhanced red nucleus responses to a site sending central programs to red nucleus. The difference between the two hypotheses, of course, is that the one concerning motor set involves the view that the response enhancement is based on short-term gating, whereas Tsukahara's hypothesis presumes that the response enhancement is due to synaptic growth.

EXPERIMENTS TO IDENTIFY BRAIN CIRCUITS FOR "SET"

The preceding section has considered two sorts of electrophysiological data — electrically evoked spikes and field potentials — that might be used to locate brain circuits underlying set-dependent changes in behaving monkeys. What, then, are some specific examples of set-dependent changes that might be elucidated by acquisition of these data? We will focus on one example, already mentioned: changes from feedback control during the set for postural stability to centrally programmed control during the set for high-velocity movement.

One may begin by recalling that EMG responses to muscle stretch are attenuated during and just before a high-velocity movement (Desmedt and Godaux, 1978a, b; 1979). This attenuation is not "all or none," but varies with movement velocity, and it has been proposed (cf. Allen and Tsukahara, 1974) that the shift from postural stability to rapid movement involves a shift of dominant control of motor cortex output by kinesthetic feedback from interpositus to central commands from dentate. A number of other putative pathways exist for both feedback and centrally programmed control of MI, but we begin by considering the hypothesis that the interpositus nucleus is one of the links underlying feedback control of MI, whereas the dentate nucleus is one of the links for centrally programmed control of MI.

As discussed in Chapter 7, electrophysiological analyses show that the cerebello-thalamocortical pathways from dentate and interpositus converge upon individual motor cortex PTNs. The intense, short-latency responses to afferent inputs in interpositus (IP) have led to the view (Allen and Tsukahara, 1974; Strick, 1983) that IP is one of the sources of inputs underlying MI responses to kinesthetic feedback. Another cerebellar output nucleus, the dentate (DEN), lacks such strong peripheral inputs, and is widely thought to be a source of inputs to MI during centrally programmed movements. These ideas lead to the hypothesis that gating may take place at some point between the cerebellar nuclei and MI with the result that feedback via IP exerts control of MI during muscle activity that utilizes feedback, whereas centrally programmed inputs via DEN are the dominant controller during the initial phases of high-velocity movement.

In our discussion of cerebello-thalamocortical electrophysiology, we considered three potential gating sites:

1. IP and DEN of cerebellum, where gating could result from inhibitory and disinhibitory inputs onto cells in these nuclei from the Purkinje cells of the cerebellar cortex.
2. VL complex of nuclei in the thalamus, where gating could result from inputs from cerebral cortex and/or from the reticular nucleus of the thalamus.
3. MI, where gating could result from corticocortical and/or nonspecific thalamic inputs.

Each of these three sites is plausible, but our further discussions do not deal with the first of the three. This omission by no means indicates that we believe gating in DEN and IP is unlikely.

GATING IN MI

Impulses evoked by stimuli restricted to either DEN or IP converge on smaller MI PTNs mainly via a trisynaptic pathway that includes one synapse in VL and two in MI. The existence of an interneuron between VL terminals in MI and smaller MI PTNs provides a site for interneuronal gating. Figure 8.6 schematically represents gating of DEN and IP control of a small MI PTN depending on motor set. In this figure it is assumed that there are MI interneurons which receive cerebellar inputs and inputs from "set cells." Examples of set cells both within MI and in areas projecting to MI were described in Chapter 6. For the purposes of the present discussion, it is not necessary to specify the location of the cell bodies of the set cells, but we stress that, in our view, the set cells are to be thought of as parts of circuits rather than as independent or autonomous elements. We assume that these set cells can control the excitability of MI interneurons that are

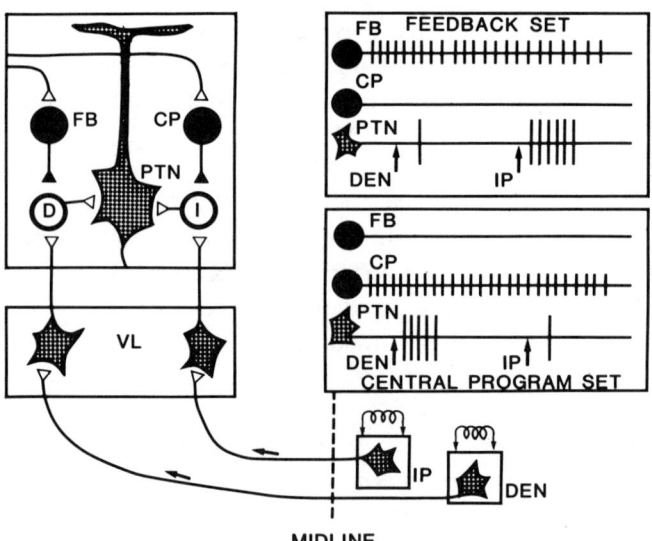

Figure 8.6. *Shift from IP control during a feedback set to DEN control during a central program set.* It is assumed that control of PTN output leading to motor behavior can be flexibly shifted from a cortical interneuron (I) relaying feedback from IP during a set for postural stability to a cortical interneuron (D) relaying a central program at the initiation of a high-velocity movement (i.e., during a set for predominant central program control). During feedback control, the hypothetical FB interneuron (part of a circuit whose discharge inhibits central program access to the PTN via D) is turned on while the hypothetical CP interneuron is silent. These two cells reverse their discharge frequencies with the shift from feedback to central program control. The hypothetical changes in the inhibitory neurons and resultant changes of PTN responses are depicted at the right.

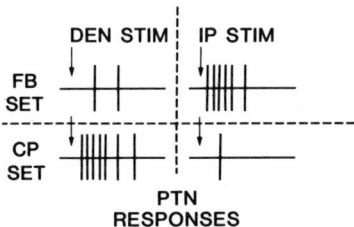

Figure 8.7. *"Double dissociation" of set-dependent responses.* If one were to record PTN impulses such as those illustrated in Figure 8.6 during the feedback (FB) set and the central program (CP) set, it might be possible to observe "double dissociation" of responses to two different input pathways as a function of the two different sets. Such a result would provide the strongest evidence for a neural switching process that is selective for particular pathways and is reflected in the activity of identified cortical cell types (see Chapter 10).

selectively impinged on by VL terminals relaying signals from DEN and IP (see Chapter 7), and that there are two groups of MI interneurons, one receiving an independent signal from DEN and the other from IP. What might the consequences be for electrically evoked MI responses during feedback as compared to centrally programmed sets? The hypothesis that DEN will control the PTN during centrally programmed sets, and that IP will control the same PTN during feedback sets would imply that the PTN response to DEN and IP should change according to the diagrams of Figure 8.7, which depicts an imaginary result in which the same PTN is activated weakly by DEN and strongly by IP during the feedback set and has a reversed response pattern during the centrally programmed set.

SET-DEPENDENT GATING OF CEREBELLAR SIGNALS IN THALAMIC NUCLEI

If one were to find a shift from IP to DEN control of MI with a change of set, then one might seek to locate the synaptic sites where the shift takes place. The change depicted in Figure 8.7 could have resulted from intracortical gating, but gating could also have occurred in VL. Figure 8.8 parallels Fig-

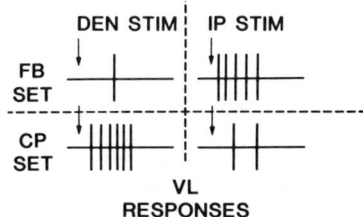

Figure 8.8. *Set-dependent gating of VL responses.* This figure parallels Figure 8.7, but now the hypothetical site of switching has been shifted from cortex to thalamus.

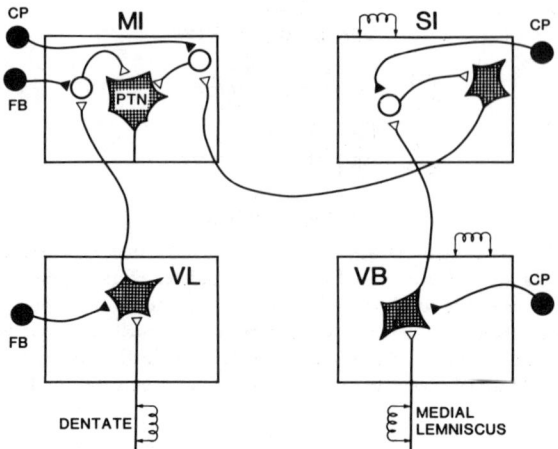

Figure 8.9. *Convergence of corticocortical and thalamocortical inputs in MI.* In addition to receiving inputs from dentate and interpositus nuclei via thalamus, motor cortex receives inputs via corticocortical fibers from the postcentral gyrus (see Chapter 7). These inputs from the postcentral gyrus may provide feedback information, and just as a set-related shift between dentate and interpositus in control of MI may be envisaged, so too one can imagine that neural switching mechanisms might allow set-dependent shifts in the effectiveness of the dentate nucleus and the postcentral gyrus in controlling output from the motor cortex. Effects on somatic sensory inputs to MI could be exerted at either thalamic or cortical levels. A way in which these possibilities could be examined is discussed in the text.

ure 8.7, but now recording is from VL rather than from MI. If gating were present in VL, it would be expected that the single cell depicted in Figure 8.8 would be more strongly activated by IP in the feedback set and more strongly activated by DEN during the centrally programmed set. According to this hypothesis, VL would not be merely a "relay," but also would serve as a switching station that could be gated according to motor set.

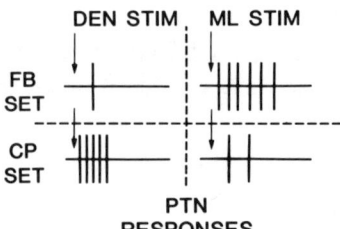

Figure 8.10. *Switching of cerebellar and lemniscal responses.* This hypothetical result would point to set-dependent gating in pathways providing central programs from dentate versus feedback from medial lemniscus. As with the Figures 8.7 and 8.8, the change cannot be due to a nonspecific alteration of excitability in the PTN itself, since at the time that one pathway becomes less effective another pathway to the same cell becomes more effective.

GATING IN CORTICOCORTICAL PATHWAYS

In addition to its thalamocortical inputs from VL, MI receives corticocortical inputs from a number of areas, including primary and secondary somatosensory areas and area 5 in the parietal lobe (see Chapter 7). Some parietal inputs might convey feedback signals to MI, and it would be reasonable to look for set-related changes of MI responses to signals from the parietal lobe. There are a number of sites at which one might deliver test stimuli to pathways conveying afferent signals to MI via the parietal lobe:

1. Medial lemniscus.
2. Ventrobasal complex of thalamus.
3. Parietal lobe.

Figure 8.9 illustrates an arrangement to test the hypothesis that there is a set-related change in the effectiveness of DEN as compared to medial lemniscus in the control of MI output via PTNs. The arrangement is somewhat more complicated than the one that compares DEN and IP, and there are more unknowns in the circuits that are schematized. But in spite of the many gaps in knowledge of the circuits, it would still be possible to compare MI PTN responses to DEN and medial lemniscus during feedback and centrally programmed sets.

Figure 8.10 shows the result that might be anticipated if the pathway conveying medial lemniscus inputs to MI were more effective during the feedback set, while the pathway conveying DEN inputs to MI were less effective. Such a result parallels the one shown for set-related changes in effects of IP and DEN, but now it is the pathway from medial lemniscus to MI, rather then the pathway from IP to MI that has become more effective during feedback control.

These examples are but a few of the many that could be readily pursued to ask questions about the short-term changes in pathway efficacy and whether such changes might *specifically* change the output of a cortical column.

Chapter 9

Behavioral Correlates of Electrically Evoked Field Potentials in the Cerebral Cortex

INTERPRETING FIELD POTENTIAL COMPONENTS

Much of our current understanding of the meaning of field potential (FP) components was obtained by Bishop and Clare (1951), who used a stimulating-recording system such as that illustrated in Figure 9.1. They concluded that of the successive FP *spikes*, all but the first represent groups of neurons firing within the cortex after successive synaptic passages, with the two *slow waves* (surface-positive and -negative) arising from two groups of neurons different from those responsible for the spikes. While attempting to relate different components of the FP to different cortical components, Bishop and Clare (1952, p. 215) recognized that there were problems:

> Having made the attempt, we conclude it is impossible by this technique to relate the source of potential components precisely to the conventional cell layers of the cortex. The reasons seem to be various. First, the cortex is a jelly, and the depth of the needle point cannot be read very accurately from a surface mark. Second, histologically the components seem to overlap cell layers, and experimentally the cells giving one component as recorded seem not to be confined to sharp boundaries and certainly not to one layer. Third, the cortex does not respond in the same pattern in successive activations. Fourth, injury probably introduces a variable. Fifth, when two potential components are coincident, such as spike and slow wave underlying it, there is no obvious criterion by which they can be separately measured. Finally, in a conducting medium the potential led off is not limited to its locus of origin, but is a distributed IR drop of potential in the field about active elements. In spite of these uncertainties some information can be elicited.

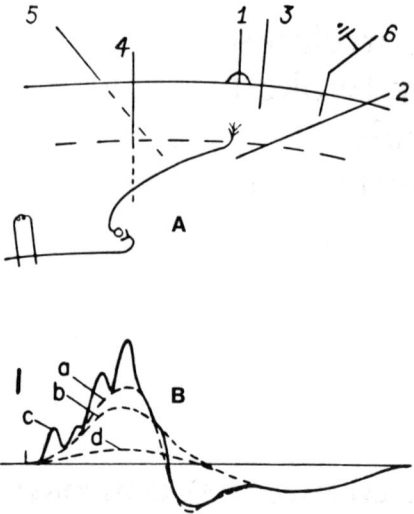

Figure 9.1. *Electrode placements to record cortical responses.* A: Electrode 1 is on the surface of the visual cortex; 2 is in white matter below the cortex and may be in contact with thalamocortical fibers; 3 is inserted into the cortex close to 1; 4 is inserted into the lateral geniculate nucleus or its fibers; 5 is a reference electrode in subcortical white matter of the temporal region; 6 is grounded. B: When the optic nerve is stimulated and recording is from point 1, a wave form such as that shown is recorded. Potential wave c was identified as arising from incoming geniculocortical fibers, while the remaining components of the response were thought to arise from intracortical elements. (From Bishop and Clare, 1951 and 1953)

MODERN INTERPRETATIONS OF FIELD POTENTIALS

Bishop's conclusion that it is impossible to relate the FP components to individual cell layers of the cortex has been supported by subsequent work, summarized by Bindman and Lippold (1981), who stated that three factors — the distribution of charges over a single cell, the possibility that individual sinks may travel, and the overlap of dendrites and cell bodies over a considerable depth of cortex — made it impossible to conclude a great deal about the anatomical site of a potential sink by finding the position at which a mass potential wave reverses in polarity.

Given these considerations, it would seem reasonable to conclude that the depth profile and phase-reversal point of evoked potentials cannot be linked directly to defined cell types. These complexities, as well as the lack of a clear theoretical relationship between FPs and single-unit activity (Amassian et al., 1964), prevent definite predictions; but it is still possible that a relationship may exist between single-unit activity and FPs. From the standpoint of practical experimental techniques, such empirically determined relationships may be useful. For example, let us suppose that we record FPs and single-unit discharges in layers III and V of cerebral cortex and discover certain reliable relations between the amplitudes and temporal characteristics of the FP components and the associated unit discharges. Once established, these relations may then provide clues that could be followed up by the much more considerable efforts involved in obtaining large samples of single units. If nothing else, this approach might be helpful in determining the laminae of the electrode tip. As we point out in Chapter 10, information about the laminae of electrode recordings might be useful in an analysis of cortical information processing. This approach has been used by Ferster and Lindström (1983), who used FPs along with antidromic techniques (see Chapter 10) to identify different layers of the striate visual cortex.

To summarize what has been said so far about cerebral cortical FPs: they may provide clues as to sites of changed pathway efficacy; they are useful because of the ease and speed with which they can be recorded (compared especially to single-unit methods discussed in Chapters 6 and 10); they can be recorded with presently available techniques. Given these advantages, it would seem worthwhile to proceed with further consideration of cerebral cortical field potentials.

ELECTRICALLY EVOKED FIELD POTENTIALS

The remainder of this chapter is devoted to showing how an ancient electrophysiological technique, the recording of electrically evoked field potentials, might be useful in experiments to detect set-related changes of pathway effectiveness. Presynaptic FPs of a bundle of axons to direct electrical stimulation follow each stimulus at fixed latency and at high repetition rate, whereas postsynaptic responses vary with behavioral state and follow less and less faithfully as the synaptic chain lengthens. These differences are of potential assistance in distinguishing between pre- and postsynaptic components of FPs.

Figure 9.2 illustrates the visual cortical FPs evoked by electrical stimulation of optic nerve or lateral geniculate fibers. The first component of the FP arises from the geniculate radiation fibers, and subsequent components are thought to arise primarily from postsynaptic intracortical elements. Responses similar to the one shown in Figure 9.2 may be recorded in MI after VL stimulation, from somatosensory cortex after stimulation of the thalamic ventrobasal complex, and from auditory cortex after media geniculate stimulation. In each case, the FP begins with a positive deflection due to activity in the incoming presynaptic thalamocortical fibers. This initial pre-

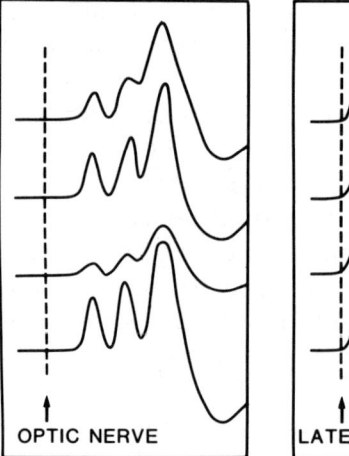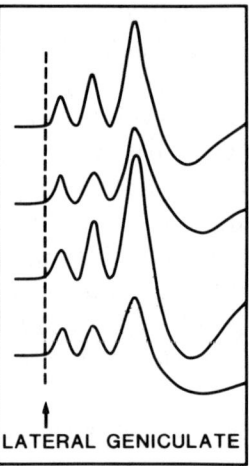

Figure 9.2. *Cortical responses to thalamic inputs.* These drawings represent visual cortex responses to electrical stimuli delivered to the optic nerve (left) or to the geniculocortical fibers (right). Note that for optic nerve stimulus, the first response component (activity in cortical terminals of geniculocortical fibers) has a longer latency at left than at right, due to synaptic delay between optic nerve fibers and lateral geniculate output. There is also more variability in the component at left, due to fluctuations in synaptic transmission within the lateral geniculate nucleus. When the stimulus is delivered to geniculocortical fibers, however, the thalamic synapse is bypassed, thereby reducing the latency of the first cortical deflection and eliminating variations of amplitude. Later components at the right still show amplitude variations, however, since such components reflect postsynaptic cortical events.

synaptic wave is virtually identical in amplitude from one trial to the next, and is almost completely unaffected by anesthesia or level of arousal. Figure 9.2 also shows cortical FP changes depending on whether the stimulus is delivered to an input to the thalamus (e.g., cerebellothalamic or optic nerve) or to the thalamocortical output fibers. With stimulation of the input there is: an increase in latency of the cortical response due to a synaptic delay in the thalamus; the lack of any components of the cortical FP that are presynaptic; and an inability to follow repetitive high-frequency stimuli with constant amplitude. With the shift of the stimulus from the lateral geniculate to the optic nerve, the amplitude of the first cortical deflection is no longer constant, but varies depending on the efficacy of the thalamic input in generating thalamic output. For example, with optic nerve stimulation, the first cortical deflection provides the same sort of information that would be provided by recording the postsynaptic component of the response to optic nerve stimulation with an electrode in the lateral geniculate nucleus. These features of electrically evoked pre- and postsynaptic responses in cortical and subcortical sites can be used to interpret results of behaviorally related FP changes by indicating relative effects of a change in behavioral state (e.g., from a motor set for movement to one of postural stability) on thalamic, as compared to cortical, pathway effectiveness. Two examples from the literature may help to clarify this use of FPs: the studies of Singer on monocular deprivation of vision and those of Steriade on sleep and arousal.

EXPERIMENTAL MODIFICATION OF FIELD POTENTIAL COMPONENTS

Singer (1977) used FPs to investigate effects of visual deprivation on the pathway from retina to visual cortex. His approach fits in well with our present discussion because it combined observations on single units and FPs. His study had three goals: to determine how the responses to electrical stimulation reflected the shift in ocular dominance; to determine the relative effects of deprivation on excitatory and inhibitory responses; and to locate the site of impaired transmission in the pathway from the deprived eye. These three goals parallel those for an experiment aimed at identifying the point in a pathway showing changes with different motor sets. Although the paradigm used by Singer was totally different from any that would be used in experiments on a short-term process such as set, his approach

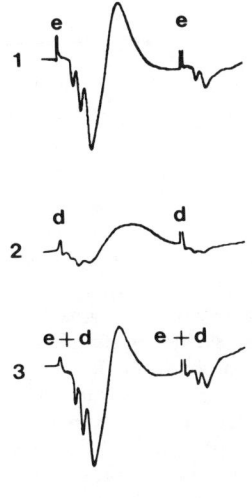

Figure 9.3. *Effects of monocular deprivation on visual cortical field potentials evoked by optic nerve stimulation.* These traces illustrate average optic nerve-evoked field potentials in the striate visual cortex ipsilateral to an eye deprived of visual experience. 1: The experienced contralateral nerve is stimulated (e) with two stimuli separated by 20 msec. 2: The deprived ipsilateral nerve is stimulated (d). 3: Both nerves are activated simultaneously with double stimuli. See Figure 9.4 and text for explanation. (From Singer, 1977)

nevertheless clarifies the way in which use of FPs could be of value in such studies. Singer showed that FPs reflected precisely the shift in ocular dominance apparent from the changes in unit receptive fields. FPs elicited from stimulation of the optic nerve of the deprived eye indicated that deprivation had affected the afferent system at the level of the lateral geniculate nucleus of the thalamus and/or at the terminal field of the thalamocortical fibers.

In Singer's experiments, it was not possible to use the response evoked from stimulation of a given optic nerve as its own control, and it was therefore necessary to compare responses evoked from the optic nerve of a deprived eye with those evoked from the optic nerve of an experienced eye. The cat optic nerve is mostly crossed, so in a normal animal the visual cortex FP is larger for stimulation of the contralateral optic nerve than for the ipsilateral nerve. In an experiment like Singer's, if the smaller FP is evoked from an *ipsilateral* deprived eye (or, more precisely, its optic nerve), some part of the amplitude difference between responses to the two nerves is due to ipsilaterality and another part to deprivation. In Singer's deprived cats, however, the magnitude of the difference (shown in Figure 9.3) was much greater than one could attribute to normal ipsilateral-contralateral

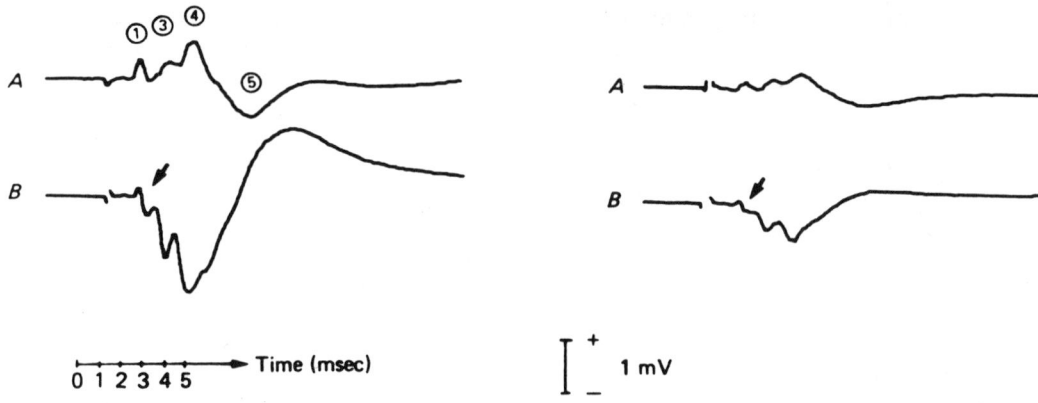

Figure 9.4. *Effects of monocular deprivation on visual cortical field potentials.* Evoked potentials recorded on the surface of the peristriate visual cortex (A) and in the subcortical white matter deep to this cortical field (B) in a monocularly deprived cat. The left traces were evoked by stimulation of the normal, ipsilateral optic nerve; the right traces were evoked by stimulation of the deprived, contralateral optic nerve. In a normal cat, the relative amplitude of ipsilateral and contralateral optic nerve-evoked potentials is the opposite of that illustrated here. The characteristic deflections in the surface potentials are thought to reflect afferent activity (1), monosynaptic activity (3), and di- or trisynaptic activity (4 and 5). The first biphasic deflections in the white matter potentials (arrows) reflect activity in the afferents as it passes by the recording site. This peak is significantly smaller in the right contralateral potential, indicating that the afferent activity to the cortex is affected by deprivation. The first peaks in the surface potentials (1) also reflect this difference. (From Mitzdorf and Singer, 1980)

differences. Whereas the true magnitude of the deprivation effect must be somewhat *less* than indicated in Figure 9.3, the true magnitude of the effect is somewhat *more* than shown in Figure 9.4, based on a study of Mitzdorf and Singer (1980) where the deprived eye was contralateral. Figure 9.4 shows that the response to the deprived (contralateral) optic nerve stimulation became smaller in spite of the fact that it would normally have been larger.

Mitzdorf and Singer followed up their observations on monocular deprivation effects on FPs with an examination of current source-densities, but in comparing the sorts of information provided by current source-density analysis and FPs, they concluded that the afferent activity is reflected more clearly in the FPs than in the current source-density profiles. Current source-density analysis might eventually be applied to studies of set-related changes in pathway effectiveness, but for the present, at least, the difficulties of precisely controlling microelectrode depth and the lack of exact knowledge of electrode orientation would complicate collection and interpretation of current source-density data. By contrast, FPs can be recorded from the cortical surface. This does not mean that one should not attempt to develop techniques for current source-density analysis in behaving animals. Indeed, the technical developments to allow this analysis may emerge along with multi-electrodes for recording at many sites simultaneously. But, as elsewhere in this monograph, we will attempt to illustrate what might be realized with presently available techniques in behaving monkeys.

FIELD POTENTIALS AND SHORT-TERM CHANGES OF BEHAVIOR

Changes of visual cortex FPs with monocular deprivation and changes of red nucleus FPs with nerve cross-innervation are thought to reflect the long-term structural changes resulting from these two sorts of manipulations. But FPs have also been shown to exhibit short-term changes, and the results of Steriade et al. (1969) on MI FPs in varying levels of arousal provide an example. Steriade et al. took advantage of different stimulating sites to determine the synaptic locus of the FP changes (Figure 9.5). As indicated in the figure, Steriade et al. found that MI responses to stimulation of cerebellothalamic fibers showed four separable components that varied with state of arousal. The first of the components corresponded to activity in the cortical terminations of VL axons, whereas the subsequent components depended on activity in intracortical cell bodies and their processes. By measuring the first component of the MI response (Figure 9.5A), one can estimate changes in effectiveness of the cerebellar inputs in generating VL outputs. On the basis of such measurements, Steriade concluded that cer-

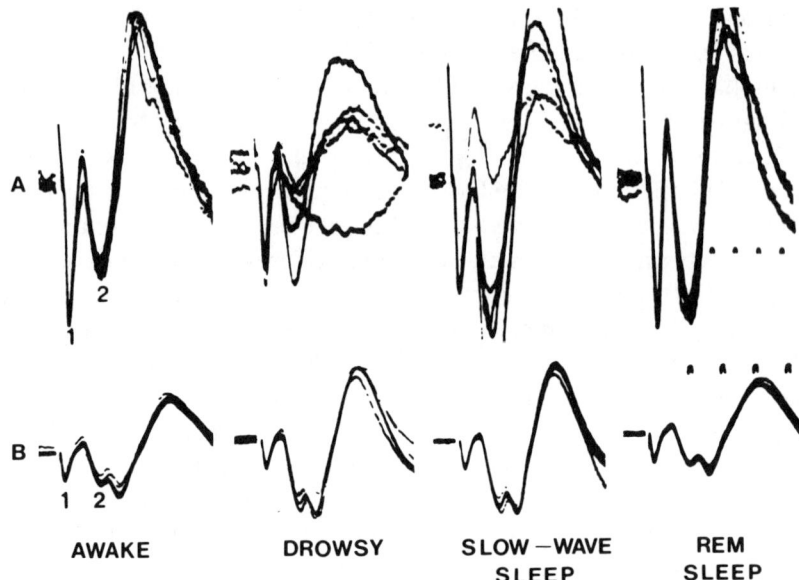

Figure 9.5. *Use of two stimulus sites to interpret field potential changes in different sleep stages.* These motor cortex responses were evoked (A) by stimulation of fibers from brachium conjunctivum (BC) or (B) by stimulation of the fibers from VL to MI. A: After BC stimulation there is variation in the amplitude of the first component of the field potential (due to fluctuations in efficacy of synaptic transmission in VL). B: In contrast, the first component is of unvarying amplitude, due to the direct stimulation of incoming thalamocortical fibers. See text for further explanation. (From Steriade et al., 1969)

ebellar inputs to VL were less effective during drowsiness and slow-wave sleep than during arousal or rapid eye movement (REM) sleep. Steriade then shifted the stimulating site from cerebellothalamic fibers to thalamocortical fibers to see if, in addition to these changes in VL, there were also arousal-related intracortical changes. This new arrangement, with its unvarying thalamocortical input, showed that drowsiness and slow-wave sleep are associated with enhanced MI FPs to a given VL input, whereas waking and REM sleep involve depressed MI FPs to the same VL inputs. Thus, Steriade et al. could detect, with FPs, relatively short-term changes in the efficacy of a pathway into the cortex and some changes (in the opposite direction) within the cortex.

USE OF ELECTRICALLY EVOKED FIELD POTENTIALS IN AN EXPERIMENT ON SET

In Chapter 8 we discussed an experiment that sought to answer the following question: Does the relative effectiveness of the pathways from the dentate nucleus to MI and from the interpositus nucleus or the medial lemniscus to MI change during sets for enhanced central program control or feedback control of MI? We now consider the logic of using FPs in one example of this sort of experiment.

The abbreviations below, a number of which have been used elsewhere in the text, serve as reminders:

IP	Interpositus nucleus of cerebellum
DEN	Dentate nucleus of cerebellum
VB	Ventrobasal complex of thalamus
VL	Ventrolateral complex of thalamus
SI	Primary somatosensory cortex
MI	Primary motor cortex
ML	Medial lemniscus
CP	Central program set
FB	Feedback set
FP	Field potential

And for short we say that

$$[\text{DEN} \to \text{MI}]_{CP}$$

means the amplitude of an MI FP evoked by dentate stimulation during a central program set, whereas

$$[\text{ML} \to \text{MI}]_{FB}$$

indicates an MI FP evoked by medial lemniscus stimulation during a feedback set, and so forth.

Let us take for our example, here, the possible contracting effects of DEN and ML inputs to the thalamus and cortex. Given that our hypothesis, following the work of Allen and Tsukahara (and others), calls for a relative increase in effect of DEN versus ML in controlling MI during centrally programmed, as compared to feedback-controlled, movement, it would follow that the result *most* supportive of our hypothesis would correspond to that shown in Figure 9.6. To express this result in the present notation:

$$[\text{ML} \to \text{MI}]_{FB} > [\text{ML} \to \text{MI}]_{CP}$$

and

$$[\text{DEN} \to \text{MI}]_{FB} < [\text{DEN} \to \text{MI}]_{CP}$$

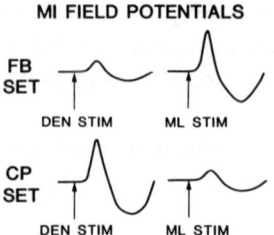

Figure 9.6. *Possible field potential changes as a function of set.* If the effectiveness of pathways generating motor cortex output were differentially changed as a function of a set for central program as compared to a set for predominent feedback control, then a motor cortex field potential recorded by a given macroelectrode might show changes such as those illustrated. Right: it is apparent that the field potential evoked by medial lemniscus (ML) is larger during the feedback set (FB) than during the central program (CP) set, whereas the reverse is true for the field potential picked up from the same electrode in response to stimulation of the dentate nucleus (DEN). This "double dissociation" of field potential changes is analogous to that hypothesized for unit discharge evoked by these sites and is discussed in the preceeding chapter (see Figure 8.10).

Similarly, in our primary example discussed in Chapter 8, one might predict that

$$[IP \to MI]_{FB} > [IP \to MI]_{CP}$$

and

$$[DEN \to MI]_{FB} < [DEN \to MI]_{CP}$$

It might be of some significance if it was also found that

$$[DEN \to MI]_{CP} > [IP \to MI]_{CP}$$

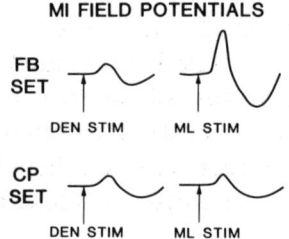

Figure 9.7. *Alternative field potential changes.* Though the hypothesized result shown in Figure 9.6 is least liable to be based upon nonspecific or generalized changes in the state of the subject, the occurrence of changes such as the one illustrated may nevertheless be of interest. This shows a possible gating effect on field potentials that might be observed if the feedback pathway via medial lemniscus is more effective during feedback-controlled movements than during centrally programmed movements, but without any concomitant overall change in the effectiveness of the pathway from dentate to motor cortex.

In addition to the possible effects discussed above and shown in Figure 9.6, other possible results warrant discussion. For example, one result might correspond to that pictured in Figure 9.7, where the feature that most interests us is

$$[ML \to MI]_{FB} > [ML \to MI]_{CP}$$

If we were to observe an enhancement of MI FPs evoked by ML stimuli during feedback control, we would list the synaptic linkages between ML and MI, and ask ourselves which of these linkages might be involved. One pathway might be ML $\to$ VB $\to$ SI $\to$ MI (see Tracey et al., 1980). Thus, we might consider the possibility of set-effects on the following FPs:

$$[ML \to VB]$$
$$[VB \to SI]$$
$$[SI \to MI]$$

To proceed to the next step, let us assume that

$$[ML \to VB]_{FB} = [ML \to VB]_{CP},$$

a result that fails to show any effect on transmission in thalamus. Of course, this result might be a false negative, since the procedure cannot distinguish between a total lack of an effect on thalamic transmission and a roughly balanced facilitation and inhibition upon different individual elements of the population. However, let us assume also that

$$[VB \to SI]_{FB} > [VB \to SI]_{CP},$$

an effect showing heightened effectiveness of the pathway from VB to SI during the feedback set. Finally, let us assume that

$$[SI \to MI]_{FB} = [SI \to MI]_{CP},$$

which is a negative result. Our use of FPs would thus have led us to suspect that there is set-related gating in SI. This finding might be followed with single-unit recording in SI in an attempt to identify different SI cell types that may play a role in gating. In the next chapter we deal with studies of impulses in such identified cell types.

Chapter 10

Behavioral Correlates of Identified Cell Types in Cerebral Cortex

In the previous chapter we discussed potential methods for examining set-related potentiation and suppression of input pathways to gross cortical regions, considered methods for assessing the efficacy of input pathways to the cortex, and speculated as to how changes in behavioral state might affect the efficacy of these inputs in the short term. However, the discussion in the previous chapter did not go into the details of how certain input pathways might be distributed to the various cortical cell types that serve to send outputs to a variety of target structures (see Chapter 5). In particular, if one finds a short-term change in the efficacy of synaptic input to a cortical region, it would be important to know how the outputs of that region are affected. To what extent is this input-output modulation specific for particular cortical outputs? Or is such modulation a general one for all outputs impinged upon by a given input? Further, in view of the difficulties in interpreting the cellular basis of evoked field potentials, single-unit confirmation of these phenomena would be a most valuable augmentation and extension of the experiments described in Chapters 8 and 9.

In this chapter, we examine some electrophysiological methods that might be used to determine whether set-related output changes are uniform or selective for one of the several types of cortical output cells. The four major techniques we consider are *fiber-tract recording*, in which the activity of single axons is monitored in a fiber tract with known origin and terminus; *antidromic axonal activation* for the purpose of identifying and locating neuronal somata that send projections into a target nucleus or fiber tract; *spike-triggered averaging* for the purpose of identifying cells evidencing postspike facilitation, either of another neuron or of muscle activity; and *laminar analysis*, consisting of a correlation of unit activity with the layer of cell somata, and inferring from this layer the most likely projection type of the cell. Although we do not refer, in detail, to the experiments described in the previous two chapters, we do conclude that there are available neurophysiological techniques by which the studies of gross cortical regions can be exploited at the level of individual cell types.

FIBER-TRACT RECORDING

The first, and technically simplest, method for determining the activity of an identified cell type is to record from the axon of that neuron in a fiber tract or bundle. For example, if one wanted to record impulses of spinal motoneurons, one could insert electrodes into a motor nerve to monitor motoneuronal action potentials, rather than inserting microelectrodes into the spinal cord. Of course, there are very few pure fiber tracts or nerves, that is, tracts which contain fibers only from one type or source of neuronal somata. Peripheral nerves, for example, typically contain both motoneuron axons and primary afferent fibers. As a result of mixing of cell types within most fiber bundles, it is not often feasible to monitor the activity of cell types by recording from fiber pathways. One pathway in which it has been possible to employ this method is the corpus callosum, a fiber tract joining the two cerebral hemispheres and containing the axons of most commissural cells. However, even here it should be noted that a certain amount of mixing of cell types is likely, because contralateral corticostriatal projections, and probably contralateral corticoclaustral projections cross through the corpus callosum. Hubel and Wiesel (1967) and Berlucchi et al. (1967) first determined the visual receptive fields of callosal axons. They found that all receptive field types (simple and complex cells of various kinds), were represented in the corpus callosum, but that the receptive fields of the cells were almost always near the vertical meridian. These investigators suggested that one of the specific roles of commissural neurons connecting the striate visual cortex (area 17) and peristriate visual cortex (area 18) of each hemisphere was to provide uniformity of the receptive fields that crossed the vertical meridian. Innocenti et al. (1972) also determined the receptive fields of commissural neurons for the somatic sensory cortex of the cat and came to strikingly different conclusions for these cortical fields.

One of the most telling weaknesses of this method, beside the rarity of "pure" pathways in the central nervous system to which it could be applied, is that, used alone, it provides little or no information concerning the location of the cell body of origin of the fiber being observed. The pathways most relevant to our discussion — the corpus callosum and pyramidal tract — contain intermingled fibers that originate from a broad extent of cerebral cortex. This intermingling is more limited in the case of the corpus callosum then in the pyramidal tract, where fibers from all areas projecting through the tract appear to distribute virtually at random (Bernard and Woolsey, 1956). Thus, additional techniques are required to determine the origin of the axon monitored. This problem can be ameliorated to some extent by stimulation techniques: one might electrically excite the area from which an axon arises and record the evoked action potentials from the isolated axon in the fiber tract. This approach, however, is also limited. It is difficult to stimulate cortical regions deep in sulci by surface electrodes, and it is prohibitively time-consuming to stimulate a large region by intracortical microstimulation. Largely for these reasons, the technique of monitoring cell types by recording from isolated axons in fiber pathways has rarely been employed. However, fiber-pathway recording has the advantage that there is no invasion of the region near the neuronal somata. This is significant, because recording at that distance means that there is none of the damage to the synaptic circuitry that must occur with extracellular recording methods that involve placement of electrodes close to cell somata. Furthermore, fiber-pathway recording eliminates the obligate electrical stimulation and possible unphysiological aftereffects of antidromic stimulation (see below).

ANTIDROMIC ACTIVATION METHODS

The second technique for identification of cell types is antidromic activation of axons in cortical pathways or target regions. Humphrey has summarized the problems and strategies of identifying such neurons (Humphrey, 1979a; see also Fuller and Schlag, 1976; Humphrey, 1968; Humphrey and Corrie, 1978; Humphrey et al., 1978; Schlag, 1978; Towe et al., 1963). Although no single criterion for identifying a cell antidromically is completely foolproof, according to Humphrey the technique will usually provide a correct decision when several criteria are used in concert. Figure 10.1 serves as the basis for Humphrey's (1979a, p. 243) discussion, which we follow closely here.

> With *intracellular* recording . . . the two modes of activation are easily distinguished by the presence of an EPSP preceding the spike during orthodromic but not antidromic activation. But with extracellular recording the following criteria must be used.
> The first criterion is that of *near-constant latency at threshold stimulus intensity* [*Figure 10.1, row 1*]. With such stimulation, an antidromically evoked spike will typically show a variation in latency (when the spike occurs) that is less than 20% of the spike duration.

However, as Humphrey points out, there are a number of problems with this criterion, including variations that may arise from changes in axon

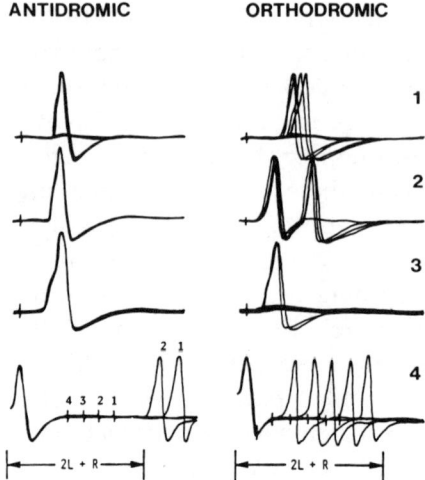

Figure 10.1. *Extracellular methods for distinguishing antidromic from orthodromic responses.* Antidromically evoked discharges have: (1) relatively constant latency at the threshold stimulus intensity; (2) a fixed latency and only single discharges at 2–3 times the threshold stimulation levels; (3) the ability to follow high stimulus repetition rates; (4) the property of collision between spontaneous and evoked spikes (with a "collision window" equal to about twice the antidromic latency (L) plus the refractory period (R). (From Humphrey, 1979a, figure 17)

excitability, in conduction velocity of the axon due to recent past history, and variation in the time between electrical stimulus and evoked axonal spike. Hence, this criterion alone is unacceptable to demonstrate antidromic activation. Humphrey continues (1979a, pp. 243–244):

> A second and related criterion is that of a *(comparatively) small decrease in latency when stimulus intensities are increased from threshold values to 2 to 3X threshold*. With such a change in intensity, an antidromically evoked spike will show typically a reduction in response latency of only a fraction of a msec. . . . With orthodromic activation, on the other hand, the decrease in latency may be on the order of several msec . . . [see Figure 10.1, row 2]. . . .
>
> A third criterion for antidromic activation is that of *faithful responding during high-frequency stimulation* (sometimes referred to as *frequency following*). During repetitive stimulation with a train of stimuli at pulse rates of 100–300/sec, an antidromically activated cell will usually respond faithfully to each pulse in the train. In contrast, an orthodromically activated cell will respond typically only to the first few stimuli in the train, before beginning to respond intermittently [see Figure 10.1 row 3]. Often, such intermittent responding occurs with stimulation rates in excess of only 10–20 pps, depending upon the complexity of the synaptic pathway.

This third criterion may not be as useful for studies of awake primates, and it should be kept in mind that each part of a cell — soma, initial segment, and axon — may have different refractory periods, and small-diameter axons may have very long refractory periods. Further, some orthodromic, monosynaptic pathways are very secure and would falsely appear to be antidromic by both the second and third tests.

According to Humphrey (1979a, p. 244):

> A fourth and perhaps the best single criterion for concluding that a unit is activated antidromically is the presence of a *collision between spontaneously occurring and evoked spikes*, when the two are separated by an appropriate interval (cf. . . . Schlag 1978). Following the occurrence of a spontaneous spike at the cell's soma, there will be an interval $I = 2L + r$ [where I = that interval, L = the antidromic latency and r = the refractory period of the axon] during which stimulation of the cell's axon will fail to produce a spike that reaches the (somatic) recording electrode. This interval is termed the *collision interval*.

By using the spike-collision method in conjunction with at least one of the additional methods described above, it would appear possible to identify cell types by antidromic activation with a high degree of certainty.

APPLICATIONS OF ANTIDROMIC IDENTIFICATION METHODS

The technique of using antidromic activation to study the properties of specific cell types has been used by many investigators of pyramidal tract neurons in precentral motor cortex. But perhaps the most illuminating examples come from the visual system. Palmer and Rosenquist (1974) set a standard when they investigated the properties of corticotectal cells in the striate visual cortex of the cat. Astonishingly, they not only found that corticotectal cells, as a population, had receptive fields different from those of other cell types, but they discovered that these cells responded with properties different from any previously observed cells. The corticotectal cells they studied were a special subclass of complex cells, one without clear summation of the response with increasing stimulus length. That is, small slits of light amounting to only about 5% of the cells' receptive fields gave approximately equal responses to slits that included the entire receptive field. These receptive fields resemble those in the superior colliculus. This correlation, along with studies in which many of the properties of tectal cells change in the absence of the striate cortex (due to ablation), suggest that cortical inputs are extremely important to the functions of the superior colliculus, at least in cats. They appear to confer upon the cat optic tectum the properties of binocularity and direction selectivity (Palmer and Rosenquist, 1974).

For our purpose, however, it is more important to note that in Palmer and Rosenquist's study, all corticotectal cells were found to be in layer V and differ in their receptive-field properties from visual receptive fields observed in the corpus callosum (and therefore are usually assumed to arise from layers II and III of area 17). Their study, then, corroborates the concept of parallel, at least partially independent, output pathways from a cortical region that convey specialized information to separate target regions (see Chapter 5). A similar study of corticopontine cells in peristriate cortex (area 18) of the cat by Gibson et al. (1978) also showed that receptive fields of corticofugal cells differed from those of the general population. The average receptive field was about eightfold greater in size than the general neuronal sample, responded much better to multiple light spots, and more selectively for the orientation of stimuli. As with the corticotectal cells examined by Palmer and Rosenquist, the corticopontine cells studied by Gibson et al. showed many of the same properties as neurons in the axonal target region, in this case the basilar pontine nuclei.

These methods have rarely been applied to the areas of cortex most directly linked to the control of voluntary movement. The studies have not often extended beyond a comparison of pyramidal tract neurons (identified by antidromic activation) and nearby cells failing to send axons through the pyramidal tract. Exceptions for the somatic sensory fields are studies by Atkinson et al. (1974), who reported different receptive-field properties in a variety of corticofugal pathways from an area of somatic sensory cortex of the cat; by Kassel (1982), who has shown that corticotectal cells from the first somatic sensory cortex of the rat have receptive fields similar to those neurons in its target zone; and by Zarzecki et al. (1983) who compared inputs to corticocortical neurons and PTNs in the cat somatic sensory cortex.

In the motor cortex, Fromm et al. (1981) reported that neurons in the primary motor cortex, with different axonal targets, can have markedly different patterns of activity, even when simultaneously recorded with an extracellular electrode. One can see in Figure 10.2 a clear difference in the activity patterns of the corticorubral neuron (CRN) and pyramidal tract neurons (PTN), though these cells were recorded in the same penetration and were within a few hundred micrometers of each other. These findings support the concept that the corticospinal and corticorubrospinal systems represent two parallel routes available to the motor cortex for control of limb motoneurons in the monkey. As Fromm et al. (1981, p. 290) conclude:

> The differences between motor cortex CRNs and PTNs (and partly also PT-CRNs) concerning time course of discharge with respect to the onset of voluntary movement, relation to steady state forces and to angular joint position, short-latency reflex responses to passive limb displacements, and background activity

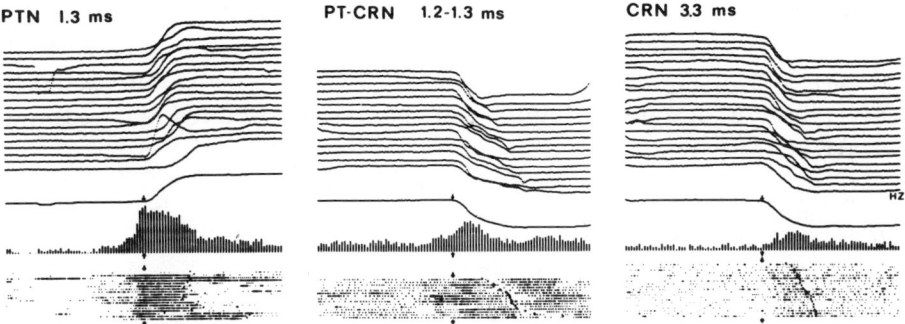

Figure 10.2. *Three types of antidromically identified primary motor cortex neurons recorded in the same penetration.* The pyramidal tract neuron (PTN) at left discharged intensely prior to movement onset, whereas the corticorubral neuron (CRN) at right exhibited no change prior to movement onset, but instead showed a slight gradual buildup of activity during movement. The branching PT-CRN in the center was intermediate in this respect. All traces at top represent position; displays are centered on the first detectable change of velocity. (From Fromm et al., 1981, figure 9)

were so clear that the assumption of a fundamentally different role for the CRN and PTN outputs from motor cortex appears to be justified, even though the numbers of PT-CRNs and CRNs in our sample were relatively small. In fact, the corticorubral projection neurons seem to have more features in common with their target RNNs [red nucleus neurons] than with the PTNs recorded in their immediate neighborhood.

There is reason to believe, on the basis of antidromic activation methods, that each cortical column can and does direct distinct outputs to a number of its cortical and subcortical targets.

The principal advantage of the antidromic activation method is that a fair degree of accuracy can be obtained as to the location of a recorded cell in the brain, and substantial information can be obtained about its axonal projections. Additionally, as Zarzecki has applied the method, it allows the assessment of subthreshold inputs in extracellular recording. With proper caution, the conclusion that a neuron is activated antidromically can be made with a high degree of confidence. Also, the branching patterns of cells can be studied, and at least crude information can be obtained about a cell's effect on target structures by stimulating the brain tissue in the immediate vicinity of the antidromically identified cell. The disadvantages of antidromic identification are: that, like all electrophysiological techniques, it is invasive, that is, one must penetrate the tissue to be stimulated and must also damage the tissue at the recording site with a microelectrode; that it is subject to false-positive results, due to spread of stimulation current outside the intended stimulation area; and that antidromic activation of an axon will also exite the collaterals of that and many other axons in a highly unphysiological synchronous volley. The possible long-term effects of such synchronous activation are not precisely known, but should be borne in mind in interpreting the activity of cells identified by antidromic axonal activation.

ELECTRODE-SAMPLING BIAS COMPENSATION AND ANTIDROMIC METHODS

A final consideration relevant to the antidromic method concerns the bias of a microelectrode toward the larger neurons in given population. Of couse, this electrode-sampling bias is common to all microelectrode studies, whether directed toward somata or axons, but when the cells are antidromically activated, one has an estimate of the cell's axonal conduction velocity. This estimate can be used to compensate for the bias of the microelectrode in a systematic population study. As Humphrey writes (1979a, pp. 246–247):

> The single unit recording method is perhaps of greatest value when it is used in a systematic *population study* (cf. Towe, 1973; Mountcastle et al., 1975; Humphrey et al., 1978). In such a study, a sensory stimulus or a behavioral response (if an alert preparation is used) is repeated under what appear to be quasi-stationary conditions, while recordings are obtained sequentially from single neurons within the neural structure or network of interest. The study typically spans several animals, and observations are collected from a sample of neurons that the investigator feels is large enough to represent well the population under study. . . .
>
> But it is quite clear to most investigators who use the method that its validity depends strongly on the adequacy and freedom from systematic bias of the unit sampling procedures. And it is equally clear that there are a host of factors which can potentially introduce such bias, even when the investigator is experienced in the basic method and aware of their existence.

The central postulates underlying any discussion of microelectrode sampling biases and methods for correction of these biases are fourfold: that the current field due to the action potential of a large cell extends over a larger volume of neural tissue than that of a small cell; that larger somata will give rise to larger axons; that larger axons conduct action potentials at a higher velocity; and that these differences in axonal conduction velocity will be adequately estimated by the time between electrical stimulation of the axon or terminal and the occurence of an action potential at the cell body, that is, the antidromic latency. It should be noted that most of these postulates have not been established directly for cortical neurons, but are principles adapted from other neuronal systems.

One can easily accept the idea that, for cells of similar geometry, the larger cells will give rise to larger extracellular field potentials, and therefore that a microelectrode is proportionately more likely to monitor the activity of these larger cells. (The same sort of phenomenon occurs in recordings from axon bundles or fiber tracts.) Additionally, since smaller cells have smaller extracellular fields, the electrode must come closer to those cells in order to monitor their activity, and therefore one runs a proportionately greater risk of damaging the cell before acquiring substantial information about its activity (Humphrey, 1979a).

Fortunately, provided that one accumulates a sufficiently large sample, Humphrey and Corrie (1978) have developed a method for arriving at an estimate of the total, or "true," population (see also Towe and Harding, 1970). The difference in the distance over which the smallest and largest cell can be monitored is probably approximately *an order of magnitude*: 20–30 micrometers for the smaller pyramidal cells in the precentral motor cortex in contrast to 200–300 micrometers for the larger pyramidal tract neurons (Humphrey, 1979a). This order-of-magnitude variation is consistent with the range in cell body areas of corticospinal cells in the precentral motor cortex (200–1800 squared micrometers) and to that of the estimated conduction velocities of pyramidal tract neurons (6–80 meters per second).

It is possible to estimate the degree of bias against a certain cell class that one may have encountered. The function which related the biased sample to the estimated population has been determined by Humphrey (Humphrey, 1979a; Humphrey and Corrie, 1978) and has been tested by comparing the distribution of axonal conduction velocities of pyramidal tract neurons with that of the fiber spectrum and the somal diameters of corticospinal neurons (or the somal diameter of layer V pyramidal cells) (Figure 10.3). The correction factor proposed by Humphrey and Corrie (1978) — which is a factor of $v^{-3/12}$ where v is axonal conduction velocity — leads to a very close correspondence of the corrected conduction velocity distribution and that of axon and cell body sizes.

An example from our own work may show how this compensation method can be used with a sample of antidromically activated pyramidal tract neurons from the precentral motor cortex. Fromm, Wise, and Evarts (1984) have attempted to apply the sampling-bias compensation method suggested by Humphrey (Humphrey and Corrie, 1978) to an analysis of sensory responses of motor cortex PTNs. Three response classes are termed dynamic, static, and dynamic-static, depending on the duration and dynamic properties of their response to a maintained displacement of the limb.

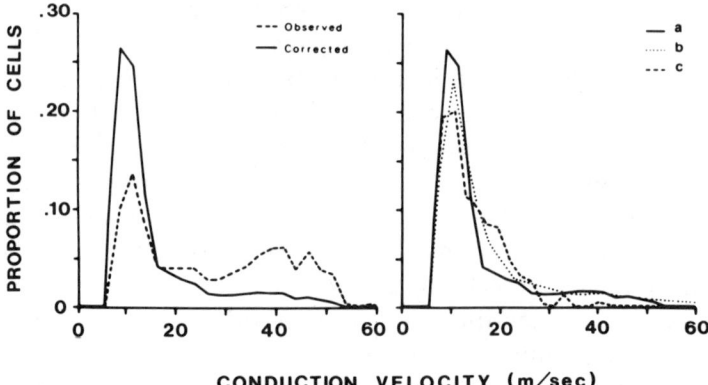

Figure 10.3. *Corrected estimates and observed samples* of axonal conduction velocities for pyramidal tract neurons. Left: experimentally observed distribution and the distribution obtained from this when a correction function for sampling bias is applied to the data (solid line). Right: a comparison of the corrected distribution (a) with the velocity distributions expected from the fiber diameters of the monkey's pyramidal tract (b) and the cell body diameters within the layer V of the cortex (c). (From Humphrey and Corrie, 1978, figure 16).

However, as described above, the smaller, slower PTNs are underrepresented in the PTN sample. These smaller PTNs are more likely to show a static response (often in conjuction with a dynamic response) than are the larger, faster PTNs (see the left and middle columns in Figure 10.4). In fact, PTNs showing a static response (including both the dynamic and dynamic-static classes) have a mean axonal conduction velocity of about 20 meters per second, whereas dynamic cells average 37 meters per second. Thus the total population of dynamic responses is likely to be overestimated, and that for static responses underestimated. When the conversion equation suggested by Humphrey is applied, there is a much more substantial, sustained (static) response among PTNs than would be appreciated from the sample population (see the right column of Figure 10.4).

In studies of the sort contemplated here, such compensation may not always be necessary. But the availability of an electrode-sampling bias compensation method may lead to an extremely significant advantage of the antidromic method. In the main, it will make it possible to identify situations in which strong biases might lead to a fallacious view of the activity of a total population. These problems are especially acute, to cite one example, in studies which attempt to compare cell activity in two cortical areas that differ in their cell-size distribution and therefore in the biases they present to the electrode. As Humphrey concludes (1979a, p. 255):

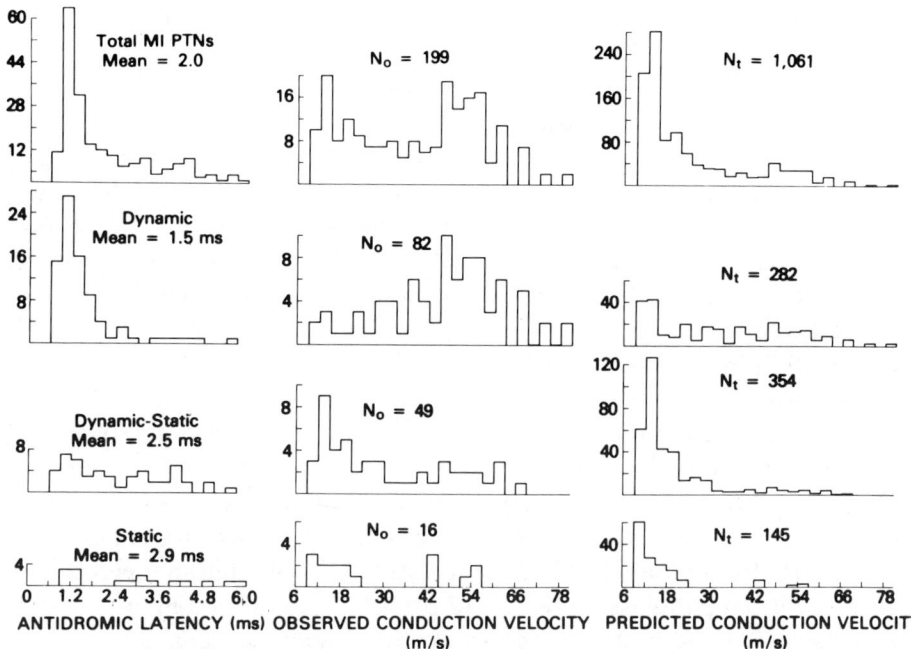

Figure 10.4. *Effect of electrode sampling bias on sensory response properties of pyramidal tract neurons (PTNs) in the precentral motor cortex.* The units were classed on the basis of the pattern of their responses to ramp displacements of the forelimb. The left column shows the mean and distribution of antidromic latencies of each of three sensory response classes: dynamic (brief responses to the stimulus), static (sustained responses to the ramp), dynamic-static (a large response that rapidly falls to a lower, sustained change in activity), and the total distribution (top). The middle column gives the estimated axonal conduction velocities of these cells and the number (N_o) observed in each class and in the total (top). The right column shows the estimated distribution and number (N_t) based on the function for electrode bias compensation proposed by Humphrey and Corrie (1978). (From Fromm, Wise and Evarts, 1984)

To what extent are such corrections necessary? Clearly, this will depend upon the nature of the cell populations under study, and the nature of the experimental questions asked. When all cells are of approximately the same size, or if functionally different groups have similar size distributions, then the effects of size-related bias should be negligible. An answer to this question will have to be formulated by each investigator, in accordance with the properties of the populations under study, and the information sought.

SPIKE-TRIGGERED AVERAGING

An alternative method has been proposed for extracellular identification of one neuronal cell type, the corticomotoneuronal (CM) cell of precentral motor cortex. This is the technique of spike-triggered averaging to elicit postspike facilitation (PSF). Fetz and Cheney (1980) have applied this method to the precentral motor cortex of the monkey. The method (illustrated in Figure 10.5) involves observing the correlation of an action potential of a motor cortex neuron with the electrical activity of muscle fibers as recorded in the electromyogram (EMG). The technique has been employed both with spike-triggered averages of muscle activity and cross-correlation analysis. The basis of the technique is probably that the cell recorded in the motor cortex contributes a fraction of the excitation to motoneurons supplying the muscle from which the EMG is recorded. This excitation, even though small, will nevertheless increase the probability of a motoneuronal spike and, therefore, the probability of an electromyographic signal synchronized with the cortical action potential, a synchronization that generally becomes apparent only when several-thousand events are averaged. Monosynaptic linkages are thought to give the largest effects, although this need not invariably be the case.

For example, in addition to an incremented EMG signal (postspike facilitation), the opposite has also been observed and termed postspike suppression (Fetz and Cheney, 1980). No known monosynaptic corticomotoneuronal linkages are inhibitory, so it is likely that this effect is mediated via interneurons. It is possible that such polysynaptic effects require averaging a much larger number of events to give rise to a clear postspike facilitation or suppression, but its existence leaves open the possibility for false-positive results with this cell type identification method. However, the spike-triggered averaging technique is clearly useful for identifying motoneuron-correlated neurons in the cortex, a proportion of which are corticomotoneuronal cells. In a detailed discussion of the method, its interpretation and statistical background, Fetz and Cheney (1980, pp. 769–770) concluded that when a relatively small number of events is sufficient to create a clear and reliable average:

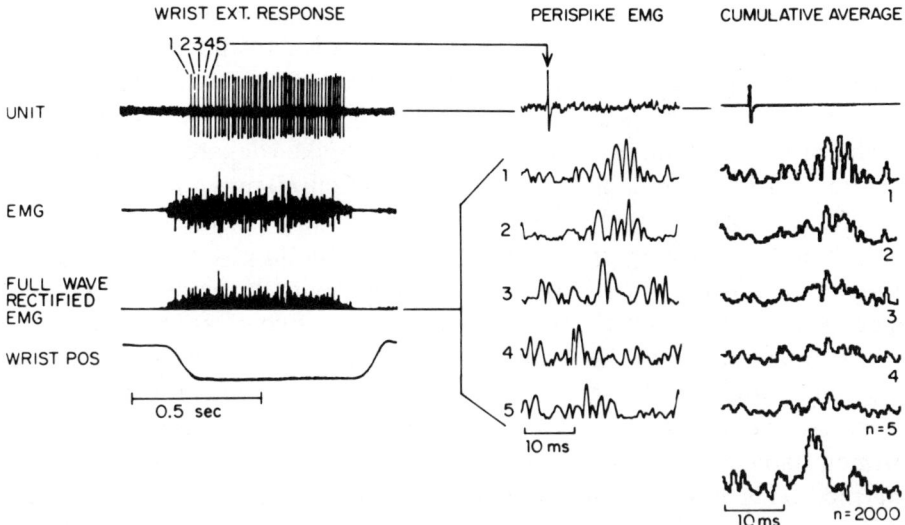

Figure 10.5. *Spike-triggered averaging technique used to detect postspike facilitation (or suppression).* The single records at left illustrate normal and rectified EMG of one agonist muscle below the record of a unit in the precentral motor cortex. The middle column illustrates the perispike-rectified EMG for the first five action potentials, and the column at right shows cumulative spike-triggered averages for the first 5 spikes, then for 2000 discharges. (From Fetz and Cheney, 1980, figure 2)

... direct connections would probably make the dominant contribution to the PSF. Irrespective of their mediation, however, the PSFs represent an empirical measure of the increased postspike firing probability of motor units correlated with the action potentials of single motor cortex cells. In that sense, the PSFs provide a more direct measure of the functional consequences of a cell's activity than could be deduced from its postsynaptic connections alone — even if such anatomical connections could ever be completely known, and could be used to predict correlational linkages. Monosynaptic CM connections probably range in effectiveness from the most potent to those too weak to produce any detectable correlogram peak. Thus, the PSF may help identify those motor cortex cells with sufficiently potent functional connections to generate enhanced postspike probability of motor unit firing. It seems likely that cortical cells with the strongest correlational linkages probably have the most direct anatomical connections. In contrast, the weaker spike-correlated effects may represent, to a debatable degree, the consequences of direct or indirect anatomical connections.

Cheney and Fetz (1980) (see also Muir and Lemon, 1983) have identified putative corticomotoneuronal cells in precentral motor cortex. They compared the motor cortex neurons that show PSF with those that do not, and found that the cortical cells which fired only during or before a movement (or isometric motor contraction), termed "phasic" cells, never showed PSF. All cells showing PSF — those putatively identified as the corticomotoneuronal cell type — showed at least some sustained activity in relation to maintained active limb displacements. Their finding further supports the notion that different cell types have distinct functional significance (see Chapter 5).

The principle advantage of the method of spike-triggered averaging is that no electrical stimulation is required in the central nervous system and, therefore, the axonal pathways of the brain should be undamaged either in the long-term by mechanical destruction or in the short-term by electrical-stimulation effects. One disadvantage is that the method does not appear to be generally applicable. It entails recording from two sites simultaneously, so it is much easier to accomplish when one of these sites is peripheral, as in electromyographic recordings. Simultaneous, two-site recording in an awake, behaving animal is a difficulty, although certainly not an insurmountable one (see Toyama et al., 1981a, b). Another disadvantage of this method is that the effects seem to be relatively small, so that thousands of events are needed to observe postspike facilitation. This is specially important if one seeks to identify cell types other than corticomotoneuronal cells. The effect, for example, of corticorubral connections on rubrospinal cells may be sufficiently slight that it would be impossible to detect with spike-triggered averaging.

These disadvantages make it unlikely that the spike-triggered averaging method for identification of cell types could find substantial application beyond the search for connections to motoneurons.

LAMINAR ANALYSIS

The final method we consider for identifying cell types is laminar analysis. In this method, the position of cell bodies in different cortical layers provides a basis for inferring cell type. Poggio et al. (1977) have used this method in a study of response properties of single neurons in visual cortex. Each unit was assigned to a cortical layer on the basis of its depth in the electrode penetration. By virtue of a given cell's laminar position, it was possible to make certain inferences as to its connectivity and its type. While, in principle, this approach could be applied to any cortical area, the inaccuracies that result from a variable amount of brain compression in chronic neurophysiological preparations and the considerable distance over which an extracellular action potential may be monitored make it difficult to apply. For work on visual cortex, however, Poggio et al. (1977) could take advantage of the presence of a layer in which high-frequency discharge was consistently encountered. This layer was later confirmed to be layer IVc. As a result,

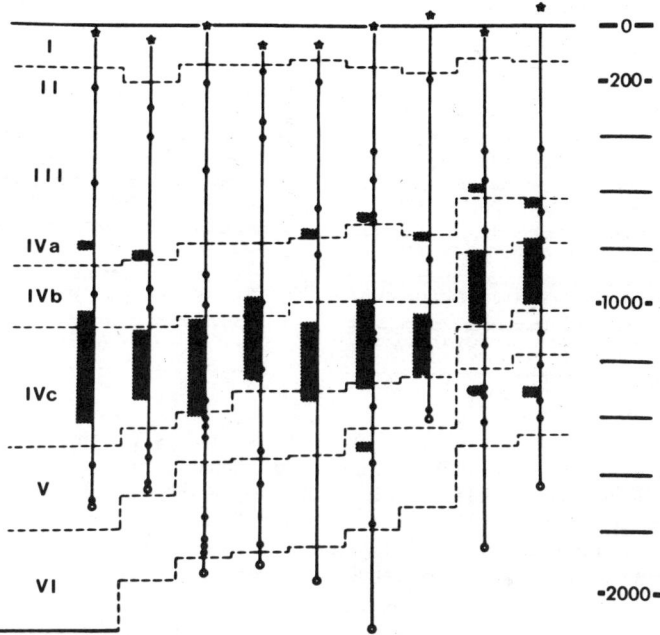

Figure 10.6. *Reconstructions of several microelctrode penetrations in the visual cortex.* A region of high spontaneous activity was localized in certain cortical layers, most notably layer IVc. Shaded areas mark the regions of sustained neuronal firing. Circles at the bottom of the histologically identified penetrations (left) mark the site of electrolytic lesions. Stars mark points at which some indication of cortical contact was obtained. (From Poggio et al., 1977, figure 4)

Poggio et al. could confidently assign neurons to layer IVc, to the infragranular layers, or to the supragranular layers (Figure 10.6). By using square-wave grating stimuli, Poggio et al. found two types of responses. Those termed "modulated" were tuned to and synchronized with the changes in brightness of the alternating light-dark gratings. "Fusional" cells also showed some tuning for the spatial frequency of the stimuli, but responded without substantial temporal modulation in relation to the changes in brightness of the moving stimulus (Figure 10.7). It was found that the fusional neurons were most common in the superficial layers of the striate visual cortex (layers II–IVb), whereas the modulated neurons were most common in the infragranular layers V and VI. They related the "fusional type" of response to *form analysis,* because those cells seem to fire to the grating stimulus, regardless of its movement, and "modulated type" of response to *motion analysis,* because the cells responded in relation to the movement of the grating within the visual field. Putting these sets of conclusion together with the probability that cells in superficial layers tend to be corticocortical

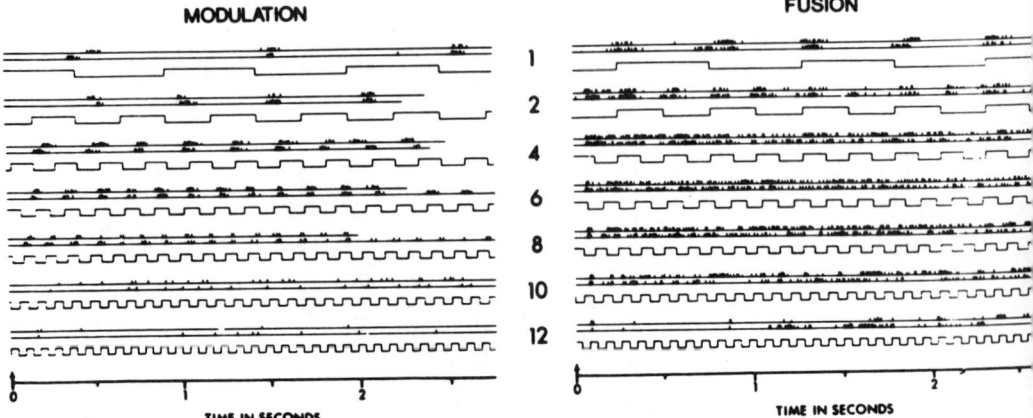

Figure 10.7. *Two categories of visual responses to moving gratings.* Two impulse trains from each cell are shown over an indication of the stimulus spatial frequency. Responses on the left show periodic modulation related to the luminance changes of the stimulus at spatial frequencies from 1 to 8 cycles per degree. Responses on the right show long bursts of firing related to luminance transitions at low spatial frequency, but show no evidence of periodic discharge in the fusion that occurs at higher spatial frequencies. (From Poggio et al., 1977, figure 5)

cells, and somata in infragranular layers tend to be corticofugal neurons, they suggested (Poggio et al., 1977, p. 1389) that their:

> ... finding that the majority of neurons in deep layers have properties appropriate for movement analyzers suggests that the pathways connecting deep striate cortex to structures associated with oculomotor behavior (superior colliculus, pulvinar) and to the lateral geniculate nucleus carry information about the motion of retinal images. Neurons in superficial layers, on the other hand, which appear to function primarily as form analyzers, send signals to cortical regions outside area 17 [*the striate visual cortex*], where it is likely that further processing occurs leading eventually to the perception of spatial structure. ...

> It may be that the local circuit processing within a vertical aggregate of cortical cells, possibly within a hypercolumn as defined by Hubel and Wiesel (1974), determines different sets of properties for neurons whose efferent projections are destined for structures having different functions.

A related approach has been taken by Gilbert and Wiesel (1981), who have identified the laminar location of striate cortex neurons. They have confirmed the results of Palmer and Rosenquist (1974), cited above, and also arrived at further inferences about the receptive field properties of the corticothalamic neurons in layer VI.

The advantage of the method of laminar analysis is that, like the spike-triggered averaging method, it does not require electrical stimulation or insertion of a stimulating electrode. Also, it is possible to collect a neuronal sample much more rapidly than with either the antidromic or spike-triggered averaging methods. Like these two other methods, the localization of the neuron in the cortex is relatively accurate, within the limits of resolution of any extracellular recording method. However, this method, as an attempt to identify specific cell types, suffers from several difficulties. First and most important, the differences between the projection targets of supragranular and infragranular cells are far from absolute. The cell bodies of corticocortical axons probably are found in all layers, including in layers VI and V (Hedreen and Yin, 1981; Jones et al., 1979a; Keller and Innocenti, 1981; Schwartz and Goldman-Rakic, 1982; Tigges et al., 1981; Wise, 1975; Zant and Strick, 1978). Thus, if one has no additional information, it is dangerous to assume that infragranular cells are corticofugal. Further, some supragranular cells and some of those in the internal granule layer (layer IV) may project to subcortical targets (Aggleton et al., 1980; Mufson et al., 1981; Royce, 1982; Swadlow and Weyand, 1981). In addition, some of the locally projecting neurons, such as basket cells (Jones, 1981a), have cell bodies as large as moderate-sized pyramidal cells and could easily be mistaken for projection neurons. But in spite of these problems, the method of laminar identification may offer clues concerning the properties of large groupings of at least tentatively identified cell types.

APPLICABILITY OF NEUROPHYSIOLOGICAL METHODS TO THE IDENTIFICATION OF CORTICAL CELL TYPES

In summary, it can be readily seen that, while use of any single method for identifying cell types in cerebral cortex has its disadvantages, each method also has advantages, and the use of several methods concurrently may ameliorate interpretational difficulties that arise from a single method used in isolation. Take, as an example, the corticorubral neuron shown in Figure 10.2. Once the activity pattern of this antidromically identified cell type has been determined, one might become concerned that an artifact was being introduced as the result of mechanical or electrical damage at the site of the antidromic stimulating electrode in the red nucleus or of recurrent effects of antidromic volleys. Then, one could record in another animal by using laminar analysis, identify the cell type on the basis of its discharge properties in a similar behavioral paradigm, and confirm that its laminar location is consistant with its identification as a corticorubral neuron. Similarly, if a particular pattern of activity were observed in antidromically identified pyramidal tract neurons in precentral motor cortex, one could overcome many interpretational questions by confirming the existence of fibers in the pyramidal tract that have the same properties, in the absence of any electrical stimulation or injury of the neuropil near the pyramidal tract somata.

By selective use of these methods to take advantage of their strengths, many cortical cell types can be identified and cortical input-output information-processing studied in relation to the behavioral contexts we have outlined in this monograph (Chapters 8 and 9): namely, short-term changes in pathway efficacy that reflect preparatory set.

Chapter 11

Summary and Speculation

SUMMARY

As noted in the Preface, this monograph has been produced under the auspices of The Neurosciences Institute. Since The Neurosciences Institute seeks, as its central goal, to assist brain scientists in the exploration of the biological basis of higher brain function at both the practical and theoretical levels, we sought to explore the potential application of currently available neurophysiological methods to study the brain circuits that underlie preparatory set. Preparatory set, as we have argued (Chapter 1), may serve as an example of one higher brain function amenable to neurophysiological analysis, in part due to the measurable motor consequences of set. The concept of set, both motor and perceptual, has been related to preparation for movement and incoming stimuli, and the behavioral flexibility evidenced by the many responses that can be triggered by a given stimulus under different conditions (Chapters 2 and 3). We have considered one of the neural sites that might be involved in this behavioral flexibility, the cerebral cortex (Chapter 4), the cell types and information-processing capabilities of the cortex (Chapter 5), and some patterns of cell activity (Chapter 6) and neural pathways (Chapter 7) that might underlie set-related behavior. By examining the finite number of neural inputs to cortical pyramidal tract neurons under a variety of conditions, it may be possible to identify the events which culminate in the swiching of neural control over motor cortex output from one input pathway to another. In an experimental situation, examining the neural activity in response to a fixed electrical stimulus may serve as a paradigm for study of the dynamic neural switching process (Chapter 8).

We used Tsukahara's experimental paradigm as a model of how one might study changes in pathway efficacy in the cerebral cortex during a variety of behavioral conditions; fir-t, to suggest the existence of postsynaptic changes in pathway efficacy (Chapter 9), then to determine how these changes may be distributed among the various cortical output elements

(Chapter 10). Tsukahara studied *long-term* changes in synaptic efficacy in the corticorubral pathway after lesion of the interpositus nucleus, cross-innervation of forelimb nerves, and classical conditioning in which an electrical stimulus delivered to the corticofugal pathway was the conditioned stimulus and limb stimulation was the unconditioned stimulus. These long-term changes can be observed by examining either the early phases of evoked field potentials, extracellularly recorded cell responses evoked by electrical stimulation of corticofugal pathways, or intracellular recording of the rise-time of excitatory postsynaptic potentials.

Our basic experimental model of *short-term* changes is conceptually similar. In its most general sense, the paradigm can be sketched as in Figure 11.1: where I_1 and I_2 are inputs to a cortical site (S_1), O is the output of that site, and A represents an agent that causes a selective potentiation of I_1 or I_2, A_u is the "ultimate agent," and A_p is the "proximate agent." The prime example we discussed (Chapters 8 and 9) is an experiment designed to examine presumed neural switching from feedback-dominated to central motor program-dominated control of motor cortex output, and testing the proposal that the shift from a set for postural stability to one for rapid movement may involve a shift of dominant control of motor cortex output from a pathway providing kinesthetic feedback to a pathway providing central commands.

In terms of our experimental model (Figure 11.1) and our prime example, I_1 is the input from the dentate nucleus and I_2 is the input from the interpositus nucleus to the precentral motor cortex (the "site," S_1). During the motor set for postural stability (in this model such a set represents A_u in the figure), the input from I_2 to the cortex is *selectively* potentiated. Further, we propose that under the condition of a set for rapid movement, I_1 is potentiated selectively. Of course, some neuronal switching elements must be involved in this short-term change in pathway efficacy, and these are indicated in

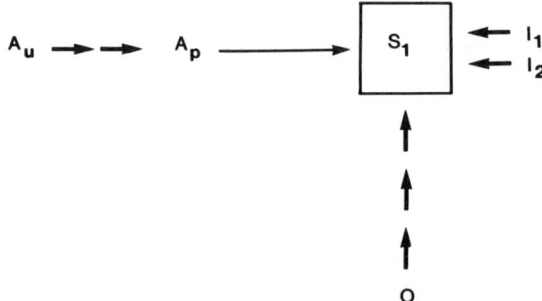

Figure 11.1. *Short-term changes in pathway efficacy.* For explanation, see text.

Figure 11.1 as A_p. The location of A_p outside of S_1 is not meant to imply that the circuits which underlie such input switching are necessarily in a different location from S_1. We do, however, propose that A_p is in a finite set of neural locations (including S_1) which directly influence S_1.

The field-potential (FP) method was proposed as a technique to determine, for a given site, whether the experimental paradigm described above appears to be valid under the behavioral circumstances of a given experiment (Chapter 9). It was concluded that, although there was a substantial potential for "false-negative" results, any observed changes in the postsynaptic evoked FP at S_1, or the presynaptic FP at the target of S_1, would suggest that the hypothesized switching occurs at S_1. The problems of interpreting the later phases of an electrically evoked field potential make it virtually impossible for this method to reveal the underlying neuronal elements that participate in this change. But, assuming that one has a procedure for locating a likely site of flexible control of input-output information processing in the cerebral cortex, available neurophysiological methods can be brought to bear on an examination of the distribution of this potentiated input among the various corticofugal cell types (Chapter 10.)

SPECULATION

Our premise has been that the preparatory set of an animal determines the selection of incoming information and preparation for the motor behavior that is to be elicited by that information. We have linked the process of set to behavioral flexibility and the rapid, efficient use of stored information (i.e., memory) and action-reward associations.

We suppose that set as a neural entity (A_p in Figure 11.1) is the dynamic, ongoing activity in a group of cells, and that these patterns form a neural representation (or, in a sense, a hypothesis) of an expected stimulus; the neural representation of a motor act which might follow such a stimulus; and the evaluation, based on experience, of whether such an act is likely to be rewarded upon receipt of the stimulus. If the motor act so represented in response to an expected stimulus is unlikely to be rewarded, a different act can be "hypothesized" in relation to the potential stimulus. The hypothesized motor act is then the new and current motor set, which is quantal in nature because the planned behavior must be real and concrete; it must be one act or another (or perhaps none at all) in response to the stimulus. If the system is in a set which indicates that stimulus A will elicit behavior B_1, and stimulus A occurs, behavior B_1 will be executed. If the operation of the system indicates that B_1 is unlikely to be successful, a set for behavior B_2 can be established and evaluated in terms of *its* likelihood of yielding a reward under the instantaneous motivational conditions. While the system is in this state, stimulus A will elicit behavior B_2. In this context, the system must have some sort of "cycle time" during which the expected signal is matched to a postulated response and evaluated, and such a system is vulnerable to failure in many ways. If the stimulus comes at a time when the system is in transition or before the evaluation has taken place, inappropriate behaviors may be elicited. On the other hand, persistence in the execution of unrewarded behaviors (perseveration) might be a different failure in this system.

To put this hypothesis on a more realistic neural basis, we need to examine what is meant by the "neural representation" of both expected stimulus (hypothesis formation) and postulated output (motor preparation). Let us suppose that these are re-entrant circuits, or loops, of the type proposed by Edelman (1978, 1981). This distributed representation resides in part in the cerebral cortex, but it is by no means confined to it. However, the activity of these re-entrant loops will be reflected by all parts of the circuit. By examining corticocortical and corticofugal projection neurons, we can monitor the status of the circuits. Certain of the circuits might be more directly connected with effector structures and circuits, and others more directly with the sensory input, but none can be considered either "motor" or "sensory" in a strict sense. A shift in the active circuits, which could be accomplished by a process that leads to some cortical projection neurons being switched off and others on, could then yield the execution of the represented behavior or a change in set, that is, a change in the motor representation from one proposed act to another or a change in perceptual set.

In such a manner, the cerebral cortex might account for the flexibility that the mammalian central nervous system offers to the individual in its interactions with a rapidly changing environment.

References

Abbs, J. H., and K. J. Cole (1982) Consideration of bulbar and suprabulbar afferent influences upon speech motor coordination and programming. In *Speech Motor Control*. S. Griller, B. Lindblom, J. Lubker, and A. Persson, eds., pp. 159–186, Pergamon, Oxford.

Aggleton, J. P., M. J. Burton, and R. E. Passingham (1980) Cortical and subcortical afferents to the amygdala of the rhesus monkey (*Macaca mulatta*). Brain Res. **190**: 347–368.

Allen, G. I., and N. Tsukahara (1974) Cerebrocerebellar communication systems. Physiol. Rev. **54**: 957–1006.

Allman, J. M., and J. H. Kaas (1975) The dorsomedial cortical visual area: A third tier area in the occipital lobe of the owl monkey (*Aotus trivirgatus*). Brain Res. **100**: 473–487.

Amassian, V. E., H. J. Waller, and J. Macy, Jr. (1964) Neural mechanism of the primary somatosensory evoked potential. Ann. N.Y. Acad. Sci. **112**: 5–32.

Amassian, V. E., and H. Weiner (1966) Monosynaptic and polysynaptic activation of pyramidal tract neurons by thalamic stimulation. In *The Thalamus*. D. P. Purpura and M. D. Yahr, eds., pp. 255–286, Columbia University Press, N.Y.

Anderson, J. A. (1977) Neural models with cognitive implications. In *Basic Processes in Reading: Perception and Comprehension*. D. La Berge and S. J. Samuels, eds., pp. 27–90, Erlbaum, Hillsdale, N.J.

Anderson, R. A., C. Asanuma, and W. M. Cowan (1982) Observations on the callosal and associational cortico-cortical connections of area 7a of the macaque monkey. Soc. Neurosci. Abstr. **8**: 210.

Aou, S., Y. Oomura, H. Nishino, T. Ono, K. Yamabe, S. K. Sikdar, T. Noda, and M. Inoue (1983) Functional heterogeneity of single neuronal activity in the monkey dorsolateral prefrontal cortex. Brain Res. **260**: 121–124.

Armstrong, E. (1983) A look at relative brain size in mammals. Neurosci. Lett. **34**: 101–104.

Arbib, M. A. (1981) Perceptual structures and distributed motor control. In *Handbook of Physiology: The Nervous System*. Vol. 2. *Motor Control*. J. M. Brookhart, V. B. Mountcastle, V. B. Brooks, and S. R. Geiger, eds., pp. 1449–1480, American Physiological Society, Bethesda, Md.

Asanuma, C., W. T. Thach, and E. G. Jones (1983) Distribution of cerebellar terminations and their relation to other afferent terminations in the ventral lateral thalamic region of the monkey. Brain Res. Rev. **5**: 237–265.

Asanuma, H. (1959) Microelectrode studies on the evoked activity of a single pyramidal tract cell in the somato-sensory area in cats. Jap. J. Physiol. **9**: 94–105.

Atkinson, D. H., J. J. Seguin, and M. Wiesendanger (1974) Organization of corticofugal neurones in somatosensory area II of the cat. J. Physiol. **236**: 663–679.

REFERENCES

Baker, J., A. Gibson, G. Mower, F. Robinson, and M. Glickstein (1983) Cat visual corticopontine cells project to the superior colliculus. Brain Res. **265**: 227–232.

Barlow, H. B. (1978) Theories of cortical function and measurements of its performance. In *Neurobiological Bases of Learning and Memory*. Y. Tsukada and B. W. Agranoff, eds., pp. 78–98, Wiley, N.Y.

Beaton, R., and J. M. Miller (1975) Single cell activity in the auditory cortex of the unanesthetized, behaving monkey: Correlation with stimulus controlled behavior. Brain Res. **100**: 543–562.

Benevento, L. A., and K. Yoshida (1981) The afferent and efferent organization of the lateral geniculo-prestriate pathways in the macaque monkey. J. Comp. Neurol. **203**: 455–474.

Benson, D. A., and R. D. Hienz (1978) Single-unit activity in the auditory cortex of monkeys selectively attending left vs. right ear stimuli. Brain Res. **159**: 307–320.

Berlucchi, G., M. S. Gazzaniga, and G. Rizzolatti (1967) Microelectrode analysis of transfer of visual information by the corpus callosum. Arch. Ital. Biol. **105**: 583–596.

Berlyne, D. I. (1969) The development of the concept of attention in psychology. In *Attention in Neurophysiology*. C. R. Evans and T. B. Mulholland, eds., pp. 1–26, Butterworths, London.

Bernard, J. W., and C. N. Woolsey (1956) Study of localization in corticospinal tracts of monkey and rat. J. Comp. Neurol. **105**: 25–50.

Bernstein, N. (1967) *The Coordination and Regulation of Movement*. Pergamon, Oxford.

Berson, D. M., and A. M. Graybiel (1978) Parallel thalamic zones in the LP-pulvinar complex of the cat identified by their afferent and efferent connections. Brain Res. **147**: 139–148.

Bindman, L., and O. Lippold (1981) *The Neurophysiology of the Cerebral Cortex*. University of Texas Press, Austin.

Bishop, G. H., and M. C. Clare (1951) Radiation path from geniculate to optic cortex in cat. J. Neurophysiol. **14**: 495–505.

Bishop, G. H., and M. C. Clare (1952) Sites of origin of electric potentials in striate cortex. J. Neurophysiol. **15**: 201–220.

Bishop, G. H., and M. C. Clare (1953) Responses of cortex to direct electrical stimuli applied at different depths. J. Neurophysiol. **16**: 1–19.

Boring, Edwin G. (1957) *A History of Experimental Psychology*. Second edition, Appleton-Century-Crofts, N.Y.

Broadbent, D. E. (1958) *Perception and Communication*. Pergamon, N.Y.

Broadbent, D. E. (1970) Stimulus set and response set: Two kinds of selective attention. In *Attention: Contemporary Theory and Analysis*. D. I. Mostofsky, ed., pp. 51–60, Appleton-Century-Crofts, N.Y.

Brooks, V. B., and W. T. Thach (1981) Cerebellar control of posture and movement. In *Handbook of Physiology: The Nervous System*. Vol. 2. *Motor Control*. J. M. Brookhart, V. B. Mountcastle, V. B. Brooks, and S. R. Geiger, eds., pp. 877–946, American Physiological Society, Bethesda, Md.

Bruce, C. J., and M. E. Goldberg (1981) Frontal eye fields on monkey: Classification of neurons discharging before saccades. Soc. Neurosci. Abstr. **7**: 131.

Bullier, J., and G. H. Henry (1979a) Ordinal position of neurons in cat striate cortex. J. Neurophysiol. **42**: 1251–1263.

Bullier, J., and G. H. Henry (1979b) Neural path taken by afferent streams in striate cortex in the cat. J. Neurophysiol. **42**: 1264–1270.

Bullier, J., and G. H. Henry (1979c) Laminar distribution of first-order neurons and afferent terminals in cat striate cortex. J. Neurophysiol. **42**: 1271–1281.

Bushnell, M. C., M. E. Goldberg, and D. L. Robinson (1981) Behavioral enhancement of visual responses in monkey cerebral cortex. I. Modulation in posterior parietal cortex related to selective visual attention. J. Neurophysiol. **46**: 755–772.

Catsman-Berrevoets, C. E., and H. G. J. M. Kuypers (1981) A search for corticospinal collaterals to thalamus and mesencephalon by means of multiple retrograde fluorescent tracers in cat and rat. Brain Res. **218**: 15–33.

Catsman-Berrevoets, C. E., R. N. Lemon, C. A. Verburgh, M. Bentivoglio, and H. G. J. M. Kuypers (1980) Absence of callosal collaterals derived from rat corticospinal neurons. Exp. Brain Res. **39**: 433–440.

Chavis, D. A., and D. N. Pandya (1976) Further observations on corticofrontal connections in the rhesus monkey. Brain Res. **117**: 369–386.

Cheney, P. D., and E. E. Fetz (1980) Functional classes of primate corticomotoneuronal cells and their relation to active force. J. Neurophysiol. **44**: 773–791.

Cherry, E. C. (1953) Some experiments on the recognition of speech, with one and with two ears. J. Acoust. Soc. Amer. **25**: 975–979.

Cherry, E. C., and W. K. Taylor (1954) Some further experiments upon the recognition of speech, with one and with two ears. J. Acoust. Soc. Amer. **26**: 554–559.

Clutton-Brock, T. H., and P. H. Harvey (1980) Primates, brains and ecology. J. Zool. **190**: 309–323.

Conrad, B., J. Meyer-Lohman, K. Matsunami, and V. B. Brooks (1975) Precentral unit activity following torque pulse injections into elbow movements. Brain Res. **94**: 219–236.

Cooper, L. (1981) Distributed memory in the central nervous system: Possible test of assumptions in visual cortex. In *The Organization of the Cerebral Cortex*. F. O. Schmitt, F. G. Worden, G. Adelman, and S. G. Dennis, eds., pp. 479–503, MIT Press, Cambridge.

Crago, P. E., J. C. Houk, and A. Hasan (1976) Regulatory actions of human stretch reflex. J. Neurophysiol. **39**: 925–935.

Davis, W. J. (1979) Behavioral hierarchies. Trends Neurosci. **1**: 5–7.

Davis, W. J., R. Gillette, M. P. Kovac, R. P. Croll, and E. M. Matera (1983) Organization of synaptic inputs to paracerebral feeding command interneurons of *Pleurobranchaea californica*. III. Modifications induced by experience. J. Neurophysiol. **49**: 1557–1572.

Delong, M. R., and A. P. Georgopoulos (1981) Motor functions of the basal ganglia. In *The Handbook of Physiology: The Nervous System*. Vol. 2. *Motor Control*. J. M. Brookhart, V. B. Mountcastle, V. B. Brooks, and S. R. Geiger, eds., pp. 1017–1061, American Physiological Society, Bethesda, Md.

Deniau, J. M., S. T. Kitai, J. P. Donoghue, and I. Grofova (1982) Neuronal interactions in the substantia nigra pars reticulata through axon collaterals of the projection neurons: An electrophysiological and morphological study. Exp. Brain Res. **47**: 105–113.

Deschenes, M. (1977) Dual origin of fibers projecting from motor cortex to S1 in cat. Brain Res. **132**: 159–162.

Deschenes, M., P. Landry, and M. Clercq (1982) A reanalysis of the ventrolateral input in slow and fast pyramidal tract neurons of the cat motor cortex. Neuroscience 7: 2149–2157.

Desiraju, T., and D. Purpura (1969) Synaptic convergence of cerebellar and lenticular projections to thalamus. Brain Res. 15: 544–547.

Desmedt, J. E., and E. Godaux (1978a) Ballistic contractions in fast or slow human muscles: Discharge patterns of single motor units. J. Physiol. 285: 185–196.

Desmedt, J. E., and E. Godaux (1978b) Ballistic skilled movements: Load compensation and patterning of the motor commands. In *Cerebral Motor Control in Man: Long Loop Mechanisms*. J. E. Desmedt, ed., pp. 21–55, Karger, Basel.

Desmedt, J. E., and E. Godaux (1979) Voluntary motor commands in human ballistic movements. Ann. Neurol. 5: 415–421.

Diamond, I. T. (1979) The subdivisions of neocortex: A proposal to revise the traditional view of sensory, motor, and association areas. Progr. Psychobiol. Physiol. Psychol. 8: 2–37.

Donoghue, J., and S. T. Kitai (1981) A collateral pathway to the neostriatum from corticofugal neurons of the rat sensory-motor cortex: An intracellular HRP study. J. Comp. Neurol. 201: 1–13.

Dormont, S. F., and C. Ohye (1971) Entopeduncular projections to the thalamic ventrolateral nucleus of the cat. Exp. Brain Res. 12: 254–264.

Eccles, J. C. (1981) The modular operation of the cerebral neocortex considered as the material basis of mental events. Neuroscience 6: 1839–1856.

Edelman, G. M. (1978) Group selection and phasic reentrant signaling: A theory of higher brain function. In *The Mindful Brain*. G. M. Edelman and V. B. Mountcastle, eds., pp. 51–95, MIT Press, Cambridge.

Edelman, G. M. (1981) Group selection as the basis for higher brain function. In *The Organization of Cerebral Cortex*. F. O. Schmitt, F. G. Worden, G. Adelman, and S. G. Dennis, eds., pp. 536–563, MIT Press, Cambridge.

Edelman, G. M., and L. H. Finkel (1984) Neuronal group selection in the cerebral cortex. In *Dynamic Aspects of Neocortical Function*. G. M. Edelman, W. M. Cowan, and W. E. Gall, eds., Wiley, N.Y., (in press).

Edelman, G. M., and G. N. Reeke Jr. (1982) Selective networks capable of representative transformations, limited generalizations, and associative memory. Proc. Natl. Acad. Sci. USA 79: 2091–2095.

Eisenberg, J. F. (1981) *The Mammalian Radiation. An Analysis of Trends in Evolution, Adaptation, and Behavior*. University of Chicago Press, Chicago.

Eisenberg, J. F., and D. Wilson (1978) Relative brain size and feeding strategies in the Chiroptera. Evolution 32: 740–751.

Eisenberg, J. F., and D. Wilson (1981) Relative brain size in didelphid marsupials. Amer. Nat. 118: 1–15.

Evarts, E. V. (1965) Relation of discharge frequency to conduction velocity in pyramidal tract neurons. J. Neurophysiol. 28: 216–228.

Evarts, E. V. (1981) Role of motor cortex in voluntary movements in primates. In *Handbook of Physiology: The Nervous System*. Vol 2. *Motor Control*. J. M. Brookhart, V. B. Mountcastle, V. B. Brooks, and S. R. Geiger, eds., pp. 1083–1120, American Physiological Society, Bethesda, Md.

Evarts, E. V., and C. Fromm (1977) Sensory responses in motor cortex neurons during precise motor control. Neurosci. Lett. **5**: 267–272.

Evarts, E. V., and R. Granit (1976) Relations of reflexes and intended movements. Progr. Brain Res. **44**: 1–14.

Evarts, E. V., and J. Tanji (1974) Gating of motor cortex reflexes by prior instruction. Brain Res. **71**: 479–494.

Evarts, E. V., and J. Tanji (1976) Reflex and intended responses in motor cortex pyramidal tract neurons of monkey. J. Neurophysiol. **39**: 1069–1080.

Evarts, E. V., and W. J. Vaughn (1978) Intended arm movements in response to externally produced arm displacements in man. Progr. Clin. Neurophysiol. **4**: 178–192.

Evarts, E. V., C. Fromm, J. Kröller, and V. A. Jennings (1983) Motor cortex control of finely graded forces. J. Neurophysiol. **49**: 1199–1215.

Ewer, R. F. (1968) *Ethology of Mammals*. Logos, N.Y.

Fairén, A., and F. Valverde (1980) A specialized type of neuron in the visual cortex of cat. A Golgi and electron microscope study of chandelier cells. J. Comp. Neurol. **194**: 761–779.

Ferster, D., and S. Lindström (1983) An intracellular analysis of geniculocortical connectivity in area 17 of the cat. J. Physiol. **342**: 181–215.

Fetz, E. E., and P. D. Cheney (1980) Postspike facilitation of forelimb muscle activity by primate corticomotoneuronal cells. J. Neurophysiol. **44**: 751–772.

Fischer, B., and R. Boch (1981a) Selection of visual targets activates prelunate cortical cells in trained rhesus monkey. Exp. Brain Res. **41**: 431–433.

Fischer, B., and R. Boch (1981b) Enhanced activation of neurons in prelunate cortex before visually guided saccades of trained rhesus monkeys. Exp. Brain Res. **44**: 129–137.

Fischer, B., and R. Boch (1983) Saccadic eye movements after extremely short reaction times in the monkey. Brain Res. **260**: 21–26.

Fischer, B., R. Boch, and M. Bach (1981) Stimulus versus eye movements: Comparison of neural activity in the striate and prelunate visual cortex (A17 and A19) of trained rhesus monkey. Exp. Brain Res. **43**: 69–77.

Fitts, P. M. (1951) Engineering psychology and equipment design. In *Handbook of Experimental Psychology*. S. S. Stevens, ed., pp. 1287–1340, Wiley, N.Y.

Forssberg, H. (1979) Stumbling corrective reaction: A phase dependent compensatory reaction during locomotion. J. Neurophysiol. **42**: 936–953.

Forssberg, H., S. Grillner, and J. Halbertsma (1980a) The locomotion of the low spinal cat. I. Coordination within a hindlimb. Acta Physiol. Scand. **108**: 269–281.

Forssberg, H., S. Grillner, J. Halbertsma, and S. Rossignol (1980b) The locomotion of the low spinal cat. II. Interlimb coordination. Acta Physiol. Scand. **108**: 283–295.

Fromm, C., E. V. Evarts, J. Kröller, and Y. Shinoda (1981) Activity of motor cortex and red nucleus neurons during voluntary movement. In *Brain Mechanisms and Perceptual Awareness*. O. Pompeiano and C. Ajmone Marsan, eds., pp. 269–294, Raven, N.Y.

Fromm, C., S. P. Wise, and E. V. Evarts (1984) Sensory response properties of pyramidal tract neurons in the precentral motor cortex and the postcentral gyrus of the rhesus monkey. Exp. Brain Res. (in press).

Fuller, J. H., and J. D. Schlag (1976) Determination of antidromic excitation by the collision test: Problems of interpretation. Brain Res. **112**: 283–298.

Fuster, J. M. (1973) Unit activity in the prefrontal cortex during delayed response performance: Neuronal correlates of short-term memory. J. Neurophysiol. **36**: 61–78.

Fuster, J. M., and G. E. Alexander (1971) Neuron activity related to short-term memory. Science **173**: 652–654.

Fuster, J. M., R. H. Bauer, and J. P. Jervey (1982) Cellular discharge in the dorsolateral prefrontal cortex of the monkey in cognitive tasks. Exp. Neurol. **77**: 679–694.

Fuster, J. M., and J. P. Jervey (1982) Neuronal firing in the inferotemporal cortex of the monkey in a visual memory task. J. Neurosci. **2**: 361–375.

Futami, T., Y. Shinoda, and J. Yokota (1979) Spinal axon collaterals of corticospinal neurons identified by intracellular injection of horseradish peroxidase. Brain Res. **164**: 279–284.

Gallistel, C. R. (1979) *The Organization of Action: A New Synthesis.* Erlbaum, Hillsdale, N.J.

Geshwind, N. (1965) Disconnexion syndromes in animals and man. Brain **88**: 585–644.

Geshwind, N. (1980) Some special functions of the human brain. In *Medical Physiology.* Vernon B. Mountcastle, ed., pp. 647–667, Mosby, N.Y.

Gibson, A., J. Baker, G. Mower, and M. Glickstein (1978) Corticopontine cells in area 18 of the cat. J. Neurophysiol. **41**: 484–495.

Gibson, J. (1941) A critical review of the concept of set in contemporary psychology. Psychol. Bull. **38**: 781–817.

Gilbert, C. D., and T. N. Wiesel (1981) Laminar specialization and intracortical connections in cat primary visual cortex. In *The Organization of the Cerebral Cortex.* F. O. Schmitt, F. G. Worden, G. Adelman, and S. G. Dennis, eds., pp. 164–190, MIT Press, Cambridge.

Glendenning, K. K., J. A. Hall, I. T. Diamond, and W. C. Hall (1975) The pulvinar nucleus of *Galago senegalensis.* J. Comp. Neurol. **161**: 419–458.

Glenn, L. L., J. Hada, J. P. Roy, M. Deschenes, and M. Steriade (1982) Anterograde tracer and field potential analysis of the neocortical layer I projection from nucleus ventralis medialis of the thalamus in cat. Neuroscience **8**: 1860–1877.

Godschalk, M., R. N. Lemon, H. G. T. Nijs, and H. G. J. M. Kuypers (1981) Behaviour of neurons in monkey peri-arcuate and precentral cortex before and during visually guided arm and hand movement. Exp. Brain Res. **44**: 113–116.

Goldberg, M. E., and C. J. Bruce (1981) Frontal eye fields in the monkey: Eye movements remap the effective coordinates of visual stimuli. Soc. Neurosci. Abstr. **7**: 131.

Goldberg, M. E., and M. C. Bushnell (1981) Behavioral enhancement of visual responses in monkey cerebral cortex. II. Modulation in frontal eye fields specifically related to saccades. J. Neurophysiol. **46**: 773–787.

Goldman-Rakic, P. S., and M. L. Schwartz (1982) Interdigitation of contralateral and ipsilateral columnar projections to frontal association cortex in primates. Science **216**: 755–757.

Granit, R. (1977) *The Purposive Brain.* MIT Press, Cambridge.

Graybiel, A. M. (1972) Some extrageniculate visual pathways in the cat. Invest. Ophthalmol. **11**: 322–332.

Graybiel, A. M., and D. M. Berson (1981) On the relation between transthalamic and transcortical pathways in the visual system. In *The Organization of the Cerebral Cortex.* F. O. Schmitt, F. G. Worden, G. Adelman, and S. G. Dennis, eds., pp. 285–319, MIT Press, Cambridge.

Haaxma, R., and H. G. J. M. Kuypers (1975) Intrahemispheric cortical connexions and visual guidance of hand and finger movements in the rhesus monkey. Brain **98**: 239–260.

Hammond, P. H. (1956) The influence of prior instruction to the subject on an apparently involuntary neuro-muscular response. J. Physiol. **132**: 17–18.

Hebb, D. O. (1949) *The Organization of Behavior, A Neuropsychological Theory*. Wiley, N.Y.

Hebb, D. O. (1955) Drives and the C.N.S. (conceptual nervous system). Psychol. Rev. **62**: 243–254.

Hebb, D. O. (1959) A neuropsychological theory. In *Psychology: A Study of Science*. Vol. 1. *Sensory, Perceptual and Physiological Formulations*. S. Koch, ed., pp. 622–643. McGraw-Hill, N.Y.

Hebb, D. O. (1972) *Textbook of Psychology*. Third edition, pp. 77–93, Saunders, Philadephia.

Hedreen, J. C., and T. C. T. Yin (1981) Homotopic and heterotopic callosal afferents of caudal inferior parietal lobule in *Macaca mulatta*. J. Comp. Neurol. **197**: 605–622.

Hendry, S. H., and E. G. Jones (1981) Sizes and distributions of intrinsic neurons incorporating tritiated GABA in monkey sensory-motor cortex. J. Neurosci. **1**: 390–408.

Hendry, S. H. C., E. G. Jones, and J. Graham (1979) Thalamic relay nuclei for cerebellar and certain related fiber systems in the cat. J. Comp. Neurol. **185**: 679–714.

Henneman, E. (1974) Principles governing distribution of sensory input to motor neurons. In *The Neurosciences: Third Study Program*. F. O. Schmitt and F. G. Worden, eds., pp. 281–291, MIT Press, Cambridge.

Herkenham, M. (1980) Laminar organization of thalamic projections to the rat neocortex. Science **207**: 532–535.

Herrnstein, R. J., and E. G. Boring, eds. (1965) *A Source Book in the History of Psychology*. Harvard University Press, Cambridge.

Hobson, J. A. (1980) Toward a cellular neurophysiology of the reticular formation: Conceptual and methodological milestones. In *The Reticular Formation Revisited: Specifying Function for a Nonspecific System*. J. A. Hobson and M. A. B. Brazier, eds., pp. 7–29, IBRO Monograph Series, Vol. 6, Raven, N.Y.

Hocherman, S., D. A. Benson, M. H. Goldstein, Jr., H. E. Heffner, and R. D. Hienz (1976) Evoked unit activity in auditory cortex of monkeys performing a selective attention task. Brain Res. **117**: 51–68.

Hocherman, S., A. Itzhaki, and E. Gilat (1981) The response of single units in the auditory cortex of rhesus monkeys to predicted and to unpredicted sound stimuli. Brain Res. **230**: 65–86.

Houk, J. C. (1976) An assessment of stretch reflex function. Progr. Brain Res. **44**: 303–314.

Hubel, D. H., and T. N. Wiesel (1962) Receptive fields, binocular interaction and functional architecture in the cat's visual cortex. J. Physiol. **160**: 106–154.

Hubel, D. H., and T. N. Wiesel (1967) Cortical and callosal connections concerned with the vertical meridian of visual fields in the cat. J. Neurophysiol. **30**: 1561–1573.

Hubel, D. H., and T. N. Wiesel (1974) Uniformity of monkey striate cortex: A parallel relationship between field size, scatter, and magnification factor. J. Comp. Neurol. **158**: 295–305.

Humphrey, D. R. (1968) Re-analysis of the antidromic cortical response. II. On the contribution of cell discharge and PSPs to the evoked potentials. Electroenceph. Clin. Neurophysiol. **25**: 421–442.

Humphrey, D. R. (1979a) Extracellular, single-unit recording methods. In *Electrophysiological Techniques*, pp. 199–259, Society for Neuroscience, Bethesda, Md.

Humphrey, D. R. (1979b) On the cortical control of visually directed reaching: Contributions by nonprecentral motor areas. In *Posture and Movement*. R. E. Talbot and D. R. Humphrey, eds., pp. 51–112, Raven, N.Y.

Humphrey, D. R., and W. S. Corrie (1978) Properties of pyramidal tract neuron system within a functionally defined subregion of primate motor cortex. J. Neurophysiol. **41**: 216–243.

Humphrey, D. R., W. S. Corrie, and R. R. Rietz (1978) Properties of the pyramidal tract neuron system within the precentral wrist and hand area of primate motor cortex. J. Physiol. **74**: 215–226.

Humphrey, N. K. (1970) What the frog's eye tells the monkey's brain. Brain Behav. Evol. **3**: 324–337.

Hunker, J. C., and J. H. Abbs (1982) Respiratory movement control during speech: Evidence for motor equivalence. Soc. Neurosci. Abstr. **8**: 946.

Hyvärinen, J., and A. Poranen (1978a) Movement-sensitive and direction and orientation-selective cutaneous receptive fields in the hand area of the post-central gyrus in monkeys. J. Physiol. **283**: 523–537.

Hyvärinen, J., and A. Poranen (1978b) Receptive field integration and submodality convergence in the hand area of the post-central gyrus of the alert monkey. J. Physiol. **283**: 539–556.

Hyvärinen, J., A. Poranen, and Y. Jokinen (1980) Influence of attentive behavior on neuronal responses to vibration in primary somatosensory cortex of the monkey. J. Neurophysiol. **43**: 870–882.

Innocenti, G. M., T. Manzoni, and G. Spidalieri (1972) Cutaneous receptive fields of single fibers of the corpus callosum. Brain Res. **40**: 507–512.

Jackson, J. H. (1931) *Selected Writings of John Hughlings Jackson*. J. Taylor, ed., Hodder and Stoughton, London.

James, W. (1890) *The Principles of Psychology*, Holt, N.Y.

Jerison, H. J. (1977) The theory of encephalization. In *Evolution and Lateralization of the Brain*. S. J. Dimond and D. A. Blizard, eds., pp. 146–160, N.Y. Academy of Science.

Jones, E. G. (1975) Varieties and distribution of nonpyramidal cells in the somatic sensory cortex of the squirrel monkey. J. Comp. Neurol. **160**: 205–268.

Jones, E. G. (1981a) Anatomy of cerebral cortex: Columnar input-output organization. In *The Organization of the Cerebral Cortex*. F. O. Schmitt, F. G. Worden, G. Adelman, and S. G. Dennis, eds., pp. 199–235, MIT Press, Cambridge.

Jones, E. G. (1981b) Functional subdivision and synaptic organization of the mammalian thalamus. Int. Rev. Physiol. **25**: 173–245.

Jones, E. G., H. Burton, C. B. Saper, and L. W. Swanson (1976) Midbrain, diencephalic and cortical relationships of the basal nucleus of Meynert and associated structures in primates. J. Comp. Neurol. **167**: 385–419.

Jones, E. G., J. D. Coulter, and S. H. C. Hendry (1978) Intracortical connectivity of architectonic fields in the somatic sensory, motor and parietal cortex of monkeys. J. Comp. Neurol. **181**: 291–348.

Jones, E. G., J. D. Coulter, and S. P. Wise (1979a) Commissural columns in the sensory-motor cortex in monkeys. J. Comp. Neurol. **188**: 113–135.

Jones, E. G., and T. P. S. Powell (1969) Connexions of the somatic sensory cortex of the rhesus monkey. I. Ipsilateral cortical connexions. Brain **92**: 477–502.

Jones, E. G., and T. P. S. Powell (1970) An anatomical study of converging sensory pathways within the cerebral cortex of the monkey. Brain 93: 793–820.

Jones, E. G., and R. Porter (1980) What is area 3a? Brain Res. Rev. 2: 1–43.

Jones, E. G., S. P. Wise, and J. D. Coulter (1979b) Differential thalamic relationships of sensory-motor and parietal cortical fields in monkeys. J. Comp. Neurol. 183: 833–881.

Kaas, J. H., R. J. Nelson, M. Sur, and M. M. Merzenich (1981) Organization of somatosensory cortex in primates. In *The Organization of the Cerebral Cortex*. F. O. Schmitt, F. G. Worden, G. Adelman, and S. G. Dennis, eds., pp. 238–261, MIT Press, Cambridge.

Kasamatsu, T., and P. Heggelund (1982) Single cell responses in cat visual cortex to visual stimulation during iontophoresis of noradrenaline. Exp. Brain Res. 45: 317–327.

Kassel, J. (1982) Somatotopic organization of SI corticotectal projections in rats. Brain Res. 231: 247–255.

Keller, G., and G. M. Innocenti (1981) Collosal connections of suprasylvian visual areas in the cat. Neuroscience 6: 703–712.

Kievet, J., and H. G. J. M. Kuypers (1975) Basal forebrain and hypothalamic connections to the frontal and parietal cortex in the rhesus monkey. Science 187: 660–662.

Kievit, J., and H. G. J. M. Kuypers (1977) Organization of the thalamo-cortical connections to the frontal lobe in the rhesus monkey. Exp. Brain Res. 29: 299–322.

Kitai, S. T. (1981) Electrophysiology of the corpus striatum and brain stem integrating systems. In *Handbook of Physiology: The Nervous System*. Vol. 2. *Motor Control*. J. M. Brookhart, V. B. Mountcastle, V. B. Brooks, and S. R. Geiger, eds., pp. 997–1013, American Physiological Society, Bethesda, Md.

Kojima, S., and P. S. Goldman-Rakic (1982) Delay-related activity of prefrontal neurons in rhesus monkeys performing delayed response. Brain Res. 248: 43–49.

Kojima, S., M. Matsumura, and K. Kubota (1981) Prefrontal neuron activity during delayed-response performance without imperative GO signals in the monkey. Exp. Neurol. 74: 396–407.

Kolb, B., and I. Q. Whishaw (1981) Decortication of rats in infancy or adulthood produced comparable functional losses on learned and species-typical behaviors. J. Comp. Physiol. Psych. 95: 468–483.

Komatsu, H. (1982) Prefrontal unit activity during a color discrimination task with GO and NO-GO responses in the monkey. Brain Res. 244: 269–277.

Kovac, M. P., and W. J. Davis (1977) Behavioral choice: Neural mechanisms in *Pleurobranchaea*. Science 198: 632–634.

Kovac, M. P., W. J. Davis, E. M. Matera, and R. P. Croll (1983a) Organization of synaptic inputs to paracerebral feeding command interneurons of *Pleurobranchaea californica*. I. Excitatory inputs. J. Neurophysiol. 49: 1517–1538.

Kovac, M. P., W. J. Davis, E. M. Matera, and R. P. Croll (1983b) Organization of synaptic inputs to paracerebral feeding command interneurons of *Pleurobranchaea californica*. II. Inhibitory inputs. J. Neurophysiol. 49: 1539–1556.

Krettek, J. E., and J. L. Price (1977) Projections from the amygdaloid complex to the cerebral cortex and thalamus in the rat and cat. J. Comp. Neurol. 172: 687–722.

Kubota, K., and S. Funahashi (1982) Direction-specific activities of dorsolateral prefrontal and motor cortex pyramidal tract neurons during visual tracking. J. Neurophysiol. 47: 362–376.

Kubota, K., and I. Hamada (1979) Preparatory activity of monkey pyramidal tract neurons related to quick movement onset during visual tracking performance. Brain Res. **168**: 435–439.

Kubota, K., T. Iwamoto, and H. Suzuki (1974) Visuokinetic activities of prefrontal neurons during delayed-response performance. J. Neurophysiol. **36**: 1197–1212.

Kubota, K., and H. Niki (1971) Prefrontal cortical unit activity and delayed alternation performance in monkeys. J. Neurophysiol **34**: 337–347.

Kubota, K., M. Tonoike, and A. Mikami (1980) Neuronal activity in the monkey dorsolateral prefrontal cortex during a discrimination task delay. Brain Res. **183**: 29–42.

Kuo, J.-S., and M. B. Carpenter (1973) Organization of pallidothalamic projections in the rhesus monkey. J. Comp. Neurol. **151**: 201–236.

Ladd, G. T., and R. S. Woodworth (1911) *Elements of Physiological Psychology*, pp. 213–274, Scribner's, N.Y.

Lashley, K. S. (1951) The problem of serial order in behavior. In *Cerebral Mechanisms in Behavior*. L. A. Jeffress, ed., pp. 112–136. Wiley, N.Y.

Lawrence, D. G., and D. A. Hopkins (1976) The development of motor control in the rhesus monkey. Evidence concerning the role of corticomotoneuronal connections. Brain **99**: 235–254.

Lawrence, D. G., and H. G. J. M. Kuypers (1968) The functional organization of the motor system in the monkey. I. The effects of bilateral pyramidal lesions. Brain **91**: 1–14.

Lorente de Nó R. (1949) Cerebral cortex: Architectonics, intra-cortical connections. In *Physiology of the Nervous System*. Third edition. J. F. Fulton, ed., pp. 274–301, Oxford, N.Y.

Lubker, J., and T. Gay (1982) Spatio-temporal goals: Maturational and cross-linguistic variables. In *Speech Motor Control*. S. Grillner, B. Lindblom, J. Lubker, and A. Persson, eds., pp. 205–216, Pergamon, Oxford.

Lund, J. S. (1973) Organization of neurons in the visual cortex area 17 of the monkey (*Macaca mulatta*). J. Comp. Neurol. **147**: 455–496.

Mackay, D. M., and M. F. Gardiner (1972) Two strategies of information processing. Neurosci. Res. Prog. Bull. **10**: 77–78.

Malliani, A., and D. P. Purpura (1967) Intracellular studies of the corpus striatum. II. Patterns of synaptic activities in lenticular and entopeduncular neurons. Brain Res. **6**: 341–354.

McKenna, T. M., B. L. Whitsel, and D. A. Dreyer (1982) Anterior parietal cortical topographic organization in macaque monkey: A re-evaluation. J. Neurophysiol. **48**: 289–317.

Miller, J. W., M. B. Buschmann, and L. A. Benevento (1980) Extrageniculate thalamic projections to the primary visual cortex. Brain Res. **189**: 221–227.

Mikami, A., S. I. Ito, and K. Kubota (1982) Modifications of neuron activities of the dorsolateral prefrontal cortex during extrafoveal attention. Behav. Brain Res. **5**: 219–223.

Mikami, A., and K. Kubota (1980) Inferotemporal neuron activities and color discrimination with delay. Brain Res. **182**: 65–78.

Mitzdorf, U., and W. Singer (1980) Monocular activation of visual cortex in normal and monocularly deprived cats: An analysis of evoked potentials. J. Physiol. **304**: 203–220.

Moray, N. (1970) *Attention: Selective Processes in Vision and Hearing*. Academic, N.Y.

Morrison, J. H., S. L. Foote, M. E. Molliver, F. E. Bloom, and H. G. W. Lidov (1982) Noradrenergic and serotonergic fibers innervate complementary layers in monkey primary visual cortex: An immunohistochemical study. Proc. Natl. Acad. Sci. USA **79**: 2401–2405.

Motter, B. C., and V. B. Mountcastle (1981) The functional properties of the light-sensitive neurons of the posterior parietal cortex studied in waking monkeys: Foveal sparing and opponent vector organization. J. Neurosci. **1**: 3–26.

Mountcastle, V. B. (1978) An organizing principle for cerebral function: The unit module and the distributed system. In *The Mindful Brain*. G. M. Edelman and V. B. Mountcastle, eds., pp. 7–50, MIT Press, Cambridge.

Mountcastle, V. B., J. C. Lynch, A. Georgopoulos, H. Sakata, and C. Acuna (1975) Posterior parietal association cortex of the monkey: Command functions for operations within extrapersonal space. J. Neurophysiol. **38**: 871–908.

Mowbray, G. H., and M. V. Rhodes (1959) On the reduction of choice reaction times with practice. Quart. J. Exp. Psychol. **11**: 16–23.

Muakassa, K. F., and P. L. Strick (1979) Frontal lobe inputs to primate motor cortex: Evidence for four somatotopically organized 'premotor' areas. Brain Res. **177**: 176–182.

Mufson, E. J., M. M. Mesulam, and D. N. Pandya (1981) Insular interconnections with the amygdala in the rhesus monkey. Neuroscience **6**: 1231–1248.

Muir, R. B., and R. N. Lemon (1983) Corticospinal neurons with a special role in precision grip. Brain Res. **261**: 312–316.

Nashner, L. M., and P. J. Cordo (1981) Relation of automatic postural responses and reaction-time voluntary movements of human leg muscles. Exp. Brain Res. **43**: 395–405.

Nashner, L. M., M. Woollacott, and G. Tuma (1979) Organization of rapid responses to postural and locomotor-like perturbations of standing man. Exp. Brain Res. **36**: 463–476.

Niki, H. (1974a) Prefrontal unit activity during delayed alternation in the monkey. I. Relation to direction of response. Brain Res. **68**: 185–196.

Niki, H. (1974b) Prefrontal unit activity during delayed alternation in the monkey. II. Relation to absolute versus relative direction of response. Brain Res. **68**: 197–204.

Niki, H. (1974c) Differential activity of prefrontal units during right and left delayed response trials. Brain Res. **70**: 346–349.

Niki, H., and M. Watanabe (1976a) Prefrontal unit activity and delayed response: Relation to cue location versus direction of response. Brain Res. **105**: 79–88.

Niki, H., and M. Watanabe (1976b) Cingulate unit activity and delayed response. Brain Res. **110**: 381–386.

Niki, H., and M. Watanabe (1979) Prefrontal and cingulate unit activity during timing behavior in the monkey. Brain Res. **171**: 213–224.

Oda, Y., K. Kuwa, S. Miyasaka, and N. Tsukahara (1981) Modification of rubral unit activities during classical conditioning in the cat. Proc. Jap. Acad. **57**: 402–405.

Palmer, L. A., and A. C. Rosenquist (1974) Visual receptive fields of single striate cortical units projecting to the superior colliculus in the cat. Brain Res. **67**: 27–42.

Palmer, L. A., A. C. Rosenquist, and R. J. Tusa (1978) The retinotopic organization of lateral suprasylvian visual areas in the cat. J. Comp. Neurol. **177**: 237–256.

Pandya, D. N., and H. G. J. M. Kuypers (1969) Cortico-cortical connections in the rhesus monkey. Brain Res. **13**: 13–36.

Pandya, D. N., and L. A. Vignolo (1971) Intra- and interhemispheric projections of the precentral, premotor and arcuate areas in the rhesus monkey. Brain Res. **26**: 217–233.

Pasik, T., and P. Pasik (1971) The visual world of monkeys deprived of striate cortex: Effective stimulus parameters and the importance of the accessory optic system. Vis. Res. **3**: 419–435.

Passingham, R. (1981) *The Human Primate*. Freeman, N.Y.

Patton, H. D., and V. E. Amassian (1960) The pyramidal tract: Its excitation and functions. In *The Handbook of Physiology: Neurophysiology*. H. W. Magoun, ed., pp. 837–861, American Physiological Society, Washington, D. C.

Penfield, W., and T. Rasmussen (1952) *The Cerebral Cortex of Man*. Macmillan, N.Y.

Peters, A., and J. R. Connor (1983) Bipolar cells in rat visual cortex. Anat. Rec. **205**: 153A.

Peters, A., C. C. Proskauer, and C. E. Ribak (1982) Chandelier cells in rat visual cortex. J. Comp. Neurol. **206**: 397–416.

Peters, A., and J. Regidor (1981) A reassessment of the forms of nonpyramidal neurons in area 17 of cat visual cortex. J. Comp. Neurol. **203**: 685–716.

Phillips, C. G. (1981) The human brain. Phil. Trans. Roy. Soc. Lond. (Biol.) **292**: 151–153.

Poggio, G. F., R. W. Doty, Jr., and W. H. Talbot (1977) Foveal striate cortex of behaving monkey: Single-neuron responses to square-wave gratings during fixation of gaze. J. Neurophysiol. **40**: 1369–1391.

Poranen, A., and J. Hyvärinen (1982) Effects of attention on multiunit responses to vibration in the somatosensory regions of the monkey's brain. Electroenceph. Clin. Neurophysiol. **53**: 525–537.

Postman, L. (1963) Perception and learning. In *Psychology: A Study of a Science*. Vol. 5. S. Koch, ed., pp. 30–113, McGraw-Hill, N.Y.

Purpura, D. P., and E. M. Housepian (1961) Alterations in corticospinal neuron activity associated with thalamocortical recruiting responses. Electroenceph. Clin. Neurophysiol. **13**: 365–381.

Purpura, D. P., T. Scarff, and J. G. McMurtry (1965) Intracellular study of internuclear inhibitions in ventrolateral thalamic neurons. J. Neurophysiol. **28**: 487–496.

Raczkowski, D., and I. T. Diamond (1978) Connection of striate cortex in *Galago senegalensis*. Brain Res. **144**: 383–388.

Robinson, C. J., and H. Burton (1980) Organization of somatosensory receptive fields in cortical areas 7b, retroinsular, postauditory and granular insula of *M. fascicularis* J. Comp. Neurol. **192**: 69–92.

Robinson, D. L., M. E. Goldberg, and G. B. Stanton (1978) Parietal association cortex in the primate: Sensory mechanisms and behavioral modulations. J. Neurophysiol. **41**: 910–932.

Rocha-Miranda, C. E., D. B. Bender, C. G. Gross, and M. Mishkin (1974) Visual activation of neurons in inferotemporal cortex depends on striate cortex and forebrain commissures. J. Neurophysiol. **37**: 475–491.

Rockland, K., and D. N. Pandya (1979) Laminar origins and terminations of cortical connections of the occipital lobe in the rhesus monkey. Brain Res. **179**: 3–20.

Rolls, E. T., S. J. Thorpe, and S. P. Maddison (1983) Responses of striatal neurons. I. Head of the caudate nucleus. Behav. Brain Res. **7**: 179–210.

REFERENCES

Rose, J. E., and C. N. Woolsey (1949) Organizations of the mammalian thalamus and its relationship to the cerebral cortex. Electroenceph. Clin. Neurophysiol. **1**: 391–404.

Rosenkilde, C. E., R. H. Bauer, and J. M. Fuster (1981) Single cell activity in ventral prefrontal cortex of behaving monkeys. Brain Res. **209**: 375–394.

Royce, G. J. (1982) Laminar origin of cortical neurons which project upon the caudate nucleus: A horseradish peroxidase investigation in the cat. J. Comp. Neurol. **205**: 8–29.

Ruch, T. C., and J. F. Fulton (1960) *Medical Physiology and Biophysics.* Saunders, Philadelphia.

Rustioni A., and N. L. Hayes (1981) Corticospinal tract collaterals to the dorsal column nuclei of cats. Exp. Brain Res. **43**: 237–245.

Ryan, T. A. (1970) *Intentional Behavior.* Ronald, N.Y.

Sakai, M. (1974) Prefrontal unit activity during visually guided lever pressing reaction in the monkey. Brain Res. **81**: 297–309.

Schell, G. R., and P. L. Strick (1984) The origin of thalamic inputs to the arcuate premotor and supplementary motor areas. J. Neurosci. (in press).

Schiller, P. H., and J. G. Malpeli (1977) The effect of striate cortex cooling on area 18 cells in the monkey. Brain Res. **126**: 366–369.

Schlag, J. (1978) Electrophysiological mapping techniques. In *Neuroanatomical Research Techniques*. R. T. Robertson, ed., pp. 385–406, Academic, N.Y.

Schneider, W., and R. M. Shiffrin (1977) Controlled and automatic human information processing. I. Detection, search and attention. Psychol. Rev. **84**: 1–56.

Schoenfeld, W. N., and W. W. Cumming (1963) Behavior and perception. In *Psychology: A Study of a Science*. Vol. 5. S. Koch, ed., pp. 213–252, McGraw-Hill, N.Y.

Schwartz, M. L., and P. S. Goldman-Rakic (1982) Single cortical neurones have axon collaterals to ipsilateral and contralateral cortex in fetal and adult primates. Nature **299**: 154–155.

Sherk, H. (1978) Area 18 cell responses in cat during reversible inactivation of area 17. J. Neurophysiol. **41**: 204–215.

Sherrington, C. S. (1933) *The Brain and Its Mechanism: The Rede Lecture.* Cambridge University Press, Cambridge.

Shiffrin, R. M., and W. Schneider (1977) Controlled and automatic human information processing. II. Perceptual learning, automatic attending, and a general theory. Psychol. Rev. **84**: 127–190.

Shinoda, Y., T. Futami, and M. Yamazaki (1983) Synaptic organization of the cerebello-thalamo-cerebral pathway in the cat. III. Input-output organization of single thalamocortical neurons in the ventrolateral thalamus (in preparation).

Shinoda, Y., J. Kokota, and T. Futami (1981) Divergent projection of individual corticospinal axons to motoneurons of multiple muscles in the monkey. Neurosci. Lett. **23**: 7–12.

Shinoda, Y., M. Yamazaki, and T. Futami (1982) Convergent inputs from the dentate and the interpositus nuclei to pyramidal tract neurons in the motor cortex. Neurosci. Lett. **34**: 111–115.

Shinoda, Y., P. Zarzecki, and H. Asanuma (1979) Spinal branching of pyramidal tract neurons in the monkey. Exp. Brain Res. **34**: 59–72.

Sillito, A. M. (1975) The contribution of inhibitory mechanisms to the receptive field properties of neurones in the striate cortex of the cat. J. Physiol. **250**: 305–329.

Sillito, A. M. (1979) Inhibitory mechanisms influencing complex cell orientation selectively and their modifications at high resting discharge levels. J. Physiol. **289**: 33–53.

Singer, W. (1977) Effects of monocular deprivation on excitatory and inhibitory pathways in cat striate cortex. Exp. Brain Res. **30**: 25–41.

Singer, W., F. Tretter, and M. Cynader (1975) Organization of cat striate cortex: A correlation of receptive-field properties with afferent and efferent connections. J. Neurophysiol. **38**: 1080–1098.

Sperry, R. W. (1952) Neurology and the mind-brain problem. Amer. Sci. **40**: 291–312.

Steriade, M., G. Iosif, and V. Apostol (1969) Responsiveness of thalamic and cortical motor relays during arousal and various stages of sleep. J. Neurophysiol. **32**: 251–264.

Sternberg, S., S. Monsell, R. L. Knoll, and C. E. Wright (1979) The latency and duration of rapid movement sequences: Comparisons of speech and typewriting. In *Information Processing in Motor Control and Learning*. G. E. Stelmach, ed., pp. 117–152, Academic, N.Y.

Stone, J., and B. Dreher (1973) Projection of X- and Y-cells of the cat's lateral geniculate nucleus to areas 17 and 18 of visual cortex. J. Neurophysiol. **36**: 551–567.

Strick, P. L. (1976a) Anatomical analysis of ventrolateral thalamic input in primate motor cortex. J. Neurophysiol. **39**: 1020–1031.

Strick, P. L. (1976b) Activity of ventrolateral thalamic neurons during arm movement. J. Neurophysiol. **39**: 1032–1044.

Strick, P. L. (1983) The influence of motor preparation on the response of cerebellar neurons to limb displacements. J. Neurosci. **3**: 2007–2020.

Strick, P. L., and C. C. Kim (1978) Input to primate motor cortex from posterior parietal cortex (area 5). I. Demonstration by retrograde transport. Brain Res. **157**: 325–330.

Strick, P. L., and J. B. Preston (1982) Two representations of the hand in area 4 of a primate. I. Motor output organization. J. Neurophysiol. **8**: 139–159.

Suzuki, H., and M. Azuma (1977) Prefrontal neuronal activity during gazing at a light spot in the monkey. Brain Res. **126**: 497–508.

Swadlow, H. A., and T. G. Weyand (1981) Efferent systems of the rabbit visual cortex: Laminar distribution of the cells of origin, axonal conduction velocities, and identification of axonal branches. J. Comp. Neurol. **203**: 799–822.

Swets, J. A. (1964) Central factors in auditory-frequency cell activity. In *Signal Detection and Recognition by Human Observers*. J. A. Swets, ed., pp. 565–581, Wiley, N.Y.

Swets, J. A., and S. A. Sewall (1964) Stimulus versus response uncertainty in recognition. In *Signal Detection and Recognition by Human Observers*. J. A. Swets. ed., pp. 431–446, Wiley, N. Y.

Szentàgothai, J. (1979) Local neuron circuits of the neocortex. In *The Neurosciences: Fourth Study Program*. F. O. Schmitt and F. G. Worden, eds., pp. 399–415, MIT Press, Cambridge.

Tanji, J., and E. V. Evarts (1976) Anticipatory activity of motor cortex neurons in relation to direction of an intended movement. J. Neurophysiol. **39**: 1062–1068.

Tanji, J., and K. Kurata (1982) Comparison of movement-related activity in two cortical motor areas of primates. J. Neurophysiol. **48**: 633–653.

Tanji, J., K. Taniguchi, and T. Saga (1980) Supplementary motor area: Neuronal response to motor instructions. J. Neurophysiol. **43**: 60–68.

Thorpe, S. J., E. T. Rolls, and S. Maddison (1983) The orbitofrontal cortex: Neuronal activity in the behaving monkey. Exp. Brain Res. **49**: 93–115.

Tigges, J., M. Tigges, S. Anschel, N. A. Cross, W. D. Letbetter, and R. L. McBride (1981) Areal and laminar distribution of neurons interconnecting the central visual cortical areas 17, 18, 19, and MT in squirrel monkey (*Saimiri*). J. Comp. Neurol. **202**: 539–560.

Towe, A. L. (1973) Sampling single neuron activity. In *Bioelectric Recording Techniques. Part A. Cellular Processes and Brain Potentials*. R. F. Thompson and M. M. Patterson, eds., pp. 79–93, Academic, N.Y.

Towe, A. L., and G. Harding (1970) Extracellular microelectrode sampling bias. Exp. Neurol. **29**: 366–381.

Towe, A. L., H. D. Patton, and T. T. Kennedy (1963) Properties of the pyramidal system in the cat. Exp. Neurol. **8**: 220–238.

Toyama, K., M. Kimura, and K. Tanaka (1981a) Cross-correlation analysis of interneuronal connectivity in cat visual cortex. J. Neurophysiol. **46**: 191–201.

Toyama, K., M. Kimura, and K. Tanaka (1981b) Organization of cat visual cortex as investigated by cross-correlation technique. J. Neurophysiol. **46**: 202–214.

Tracey, D. J., C. Asanuma, E. G. Jones, and R. Porter (1980) Thalamic relay to motor cortex: Afferent pathways from brain stem, cerebellum, and spinal cord in monkeys. J. Neurophysiol. **44**: 532–554.

Traub, M. M., J. C. Rothwell, and C. D. Marsden (1980) A grab reflex in the human hand. Brain **103**: 869–884.

Tsukahara, N., Y. Fujito, Y. Oda, and J. Maeda (1982) Formation of functional synapses in the adult cat red nucleus from the cerebrum following cross-innervation of forelimb flexor and extensor nerves. Exp. Brain Res. **45**: 1–12.

Tsukahara, N., Y. Oda, and T. Notsu (1981) Classical conditioning mediated by the red nucleus in the cat. J. Neurosci. **1**: 72–79.

Tusa, R. J., L. A. Palmer, and A. C. Rosenquist (1975) The retinotopic organization of the visual cortex in the cat. Soc. Neurosci. Abstr. **1**: 52.

Uno. M., N. Ozawa, and M. Yoshida (1978) The mode of pallido-thalamic transmission investigated with intracellular recording from cat thalamus. Exp. Brain Res. **33**: 493–507.

Uno, M., and M. Yoshida (1975) Monosynaptic inhibition of thalamic neurons produced by stimulation of the pallidal nucleus in cats. Brain Res. **99**: 377–380.

Uno, M., M. Yoshida, and I. Hirota (1970) The mode of cerebello-thalamic relay transmission investigated with intracellular recording from cells of the ventrolateral nucleus of cat's thalamus. Exp. Brain Res. **10**: 121–139.

Vaadia, E., Y. Gottlieb, and M. Abeles (1982) Single-unit activity related to sensorimotor association in auditory cortex of a monkey. J. Neurophysiol. **48**: 1201–1213.

Valverde, F. (1976) Aspects of cortical organization related to the geometry of neurons with intra-cortical axons. J. Neurocytol. **5**: 509–529.

Van Essen, D. C. (1979) Visual areas of the mammalian cerebral cortex. Annu. Rev. Neurosci. **2**: 227–263.

Van Essen, D. C., W. T. Newsome, and J. L. Bixby (1982) The pattern of interhemispheric connections and its relationship to extrastriate visual areas in the macaque monkey. J. Neurosci. **2**: 265–283.

Vogt, B. A., and D. N. Pandya (1978) Cortico-cortical connections of somatiac sensory cortex (areas 3, 1 and 2) in the rhesus monkey. J. Comp. Neurol. **177**: 179–192.

Warren, J. (1974) Possibly unique characteristics of learning by primates. J. Hum. Evol. **3**: 445–454.

Watanbe, M. (1981) Prefrontal unit activity during delayed conditional discriminations in the monkey. Brain Res. **225**: 51–65.

Waterhouse, B. D., and D. J. Woodward (1980) Interaction of norepinephrine with cerebrocortical activity evoked by stimulation of somatosensory afferent pathways in the rat. Exp. Neurol. **67**: 11–34.

Weinrich, M., and S. P. Wise (1982) The premotor cortex of the monkey. J. Neurosci. **2**: 1329–1345.

Whishaw, I. Q., A. J. Nonneman, and B. Kolb (1981) Environmental constraints on motor abilities used in grooming, swimming, and eating by decorticate rats. J. Comp. Physiol. Psychol. **95**: 792–804.

White, E. L. (1978) Identified neurons in mouse Sml cortex which are postsynaptic to thalamocortical axon terminals: A combined Golgi-electron microscopic and degeneration study. J. Comp. Neurol. **181**: 627–662.

White, E. L. (1979) Thalamocortical synaptic relations: A review with emphasis on the projections of specific thalamic nuclei to the primary sensory areas of the neocortex. Brain Res. Rev. **1**: 275–312.

White, E. L. (1981) Thalamocortical synaptic relations. In *The Organization of the Cerebral Cortex*. F. O. Schmitt, F. G. Worden, G. Adelman, and S. G. Dennis, eds., pp. 151–161, MIT Press, Cambridge.

Wiesendanger, M. (1981) Organization of secondary motor areas of cerebral cortex. In *Handbook of Physiology: The Nervous System*. Vol. 2. *Motor Control*. J. M. Brookhart, V. B. Mountcastle, V. B. Brooks, and S. R. Geiger, eds., pp. 1121–1147, American Physiological Society, Bethesda, Md.

Wise, S. P. (1975) The laminar organization of certain afferent and efferent fiber systems in the rat somatosensory cortex. Brain Res. **90**: 139–142.

Wise, S. P. (1984) The nonprimary motor cortex and its role in the cerebral control of movement. In *The Dymanic Aspects of Neocortical Function*. G. M. Edelman, W. M. Cowan, and E. Gall, eds., Wiley, N.Y. (in press).

Wise, S. P., M. Weinrich, and K.-H. Mauritz (1983) Motor aspects of cue-related neuronal activity in premotor cortex of the rhesus monkey. Brain Res. **260**: 301–305.

Woodworth, R. S. (1958) *Dynamics of Behavior*. Holt, Rinehart & Winston, N.Y.

Woolsey, C. N. (1958) Organization of somatic sensory and motor areas of the cerebral cortex. In *Biological and Biochemical Bases of Behavior*. H. F. Harlow and C. N. Woolsey, eds., pp. 63–81, University of Wisconsin Press, Madison.

Woolsey, C. N. (1964) Cortical localization as defined by evoked potential and electrical stimulation studies. In *Cerebral Localization and Organization*. G. Schaltenbrand and C. N. Woolsey, eds., pp. 17–26, University of Wisconsin Press, Madison.

Woolsey, C. N., T. C. Erickson, and W. E. Gilson (1979) Localization in somatic sensory and motor areas of human cerebral cortex as determined by direct recording of evoked potentials and electrical stimulation. J. Neurosurg. **51**: 476–506.

REFERENCES

Woolsey, C. N., P. H. Settlage, D. R. Meyer, W. Spencer, T. Pinto Hamuy, and A. M. Travis (1952) Patterns of localization in precentral and "supplementary" motor areas and their relationship to the concept of a premotor area. Res. Pub. Assoc. Res. Nerv. Ment. Dis. **30**: 238–264.

Wurtz, R. H., and C. W. Mohler (1976) Enhancement of visual response in monkey striate cortex and frontal eye fields. J. Neurophysiol. **39**: 766–772.

Wurtz, R. H., B. J. Richmond, and W. T. Newsome (1984) Modulation of cortical visual processing by attention, perception, and movement. In *Dynamic Aspects of Neocortical Function*. G. M. Edelman, W. M. Cowan, and E. Gall, eds., Wiley, N.Y. (in press).

Zant, J. D., and P. L. Strick (1978) The cells of origin of interhemispheric connections in the primate motor cortex. Soc. Neurosci. Abstr. **4**: 308.

Zarzecki, P., P. S. Blum, D. A. Bakker, and D. Herman (1983) Convergence of sensory inputs upon projection neurons of somatosensory cortex: Vestibular, neck, head, and forelimb inputs. Exp. Brain Res. **50**: 408–414.

Zarzecki, P., Y. Shinoda, and H. Asanuma (1978a) Projection from area 3a to the motor cortex by neurons activated from group I muscle afferents. Exp. Brain Res. **79**: 2401–2405.

Zarzecki, P., P. L. Strick, and H. Asanuma (1978b) Input to primate motor cortex from posterior parietal cortex (area 5). II. Identification by antidromic activation. Brain Res. **157**: 331–335.

Zeki, S. M. (1978) Functional specialization in the visual cortex of the rhesus monkey. Nature **274**: 423–428.

Index

Abbreviations, 43, 149
Adaptive plasticity:
    and short-term set, 131-134
    structural *vs.* dynamic, 132
Afferent inputs, open *vs.* closed-loop, 34
Animals, neural activity in, 68-69
Antidromic activation:
    advantages of, 159
    applications, 157-160
    electrode-sampling bias compensation
        and, 160-164
    methods, 155-160
Area-specific experimentation, 66
Association cortex, 42
Attention:
    concepts of, 9, 11
    and consciousness, 11
    and set, 8, 9
Auditory region, of cortex, 67
Auditory temporal cortex, 87-89
    postinstruction activity, 89
Automatic responses:
    set-dependent, 35-37
    set effects, 33-34
Averaging, spike-triggered, 153
Axons:
    antidromic activation, 5, 153
    conduction velocities, 110-111, 162
    corticothalamic, 58
    in lower brainstem, 5
    refractory periods, 157
    small-diameter, 157

Basal ganglia:
    neuronal circuitry, 124
    signal flow, 123
Basket cells, 59, 62
Behavior:
    cell type correlates, 153-170
    electrophysiology and, 127
    feeding, 68
    field potentials and, 147-148
    flexibility of, 39-41
    instruction stimulus in, 7

mechanisms of, 2-3
preparatory set and, 7-20
and pyramidal tract neuron, 1-6
set-dependent, 91-126
set-related changes, 13
short-term changes of, 147-148
volitional, 30-32
Biceps discharge, 26
Bitufted cell, 64
Black box, 65-67
    cell assembly in, 13
    mediating neural events in, 12-15
    single-unit approach, 65
    see also Cerebral cortex
Brachium conjunctivum, 110
    stimulation of, 111
Brain:
    activity levels, 13
    cell assembly, 13
    circuits, 134-135
    size of, 39-40
    states, 18
Brain waves, *see* Electrophysiology

Cat, visual cortex, 53
Cell types:
    antidromic activation, 155-160
    assembly, 13-14
    behavioral correlates of, 153-170
    in cerebral cortex, 49-64, 153-170
    cortical, 169-170
    corticomotoneuronal, 164-166
    corticopontine, 158
    corticotectal, 157-158
    defined, 58
    laminar analysis, 166-169
    mixing of, 154
    neocortical, 58-59
    neurophysiological identification of,
        169-170
    nonpyramidal, 60-61
    thalamic input, 51
Cerebellar cortex, 64
    functions, 59-60

Cerebellar cortex (*Continued*)
  synaptic organization, 59
Cerebellum:
  cortical pathways, 103-114
  dentate nucleus of, 149
  electrophysiology, 121-122
  interpositus nucleus of, 149
  set-dependent gating, 137-138
  thalamic inputs, 96-99
Cerebral cortex, 39-47
  behavioral flexibility, 39-41
  cell types in, 49-64, 153-170
  corticocortical inputs, 52-55
  experimental model, 172-173
  field potentials in, 141-151
  gating circuits, 62
  information-processing circuits in, 49-64
  kinesthetic input, 80
  motor area, 5
  movement control, 44-46
  neurophysiology, 1-3
  nonpyramidal cells, 60-61
  organization of, 41
    columnar, 49-50
    intrinsic, 59-64
  output cells, 60
  somatosensory areas, 42
  thalamic nuclei and, 13
Cerebro-cerebellar pathways, 104
Chandelier cell, 61
Conditioning:
  and nerve cross-union, 127
  in red nucleus, 128
Convergence:
  cerebello-thalamocortical, 119
  of corticocortical and thalamocortical inputs, 138
  of dentate and interpositus outputs, 120
  and occlusion, 116-117
Corneal reflex, 32
Corpus callosum, fiber-tract recording in, 154
Cortex:
  efferent cell types, 57
  frontal agranular, 78-85
  frontal granular, 69-78
  inputs to, 50-56
  microelectrode studies, 44
  motor areas of, 42
  prefrontal, 69-78
  primary motor, 1-2
  rat visual, 51
  regions, 67
  somatic sensorimotor, 42

somatosensory:
  postcentral, 43
  second, 43
  supplementary sensory, 44
Cortical fibers, dependence on, 55
Cortical fields, 42
  identification of, 43
  somatic sensorimotor, 43
Cortical lamina, 41
Cortical outputs:
  organization of, 56-58
  types of, 56-57
Cortical sheet, dimensions of, 41
Corticocortical inputs, to cerebral cortex, 52-55
Corticofugal cells, 57, 58
Corticothalamic fibers, 95
Cough reflex, 32
Cross-innervation, in red nucleus, 129
Cue(s), auditory neuronal, 76
Cueing, 76
  in signal detection, 17

Dead time, in set switching, 20
Deprivation:
  effects, 147
  monocular, 146, 147
  visual, 145-146
Dichotic listening experiments, 18
Direct response, 67
Drowsiness, cerebellar inputs during, 148
Dwell time in set switching, 20

Electrode(s), recording-type, 113
Electrode-sampling bias, 160-164
  degree of, 161
  effects of, 163
Electrophysiology, 103, 143
  and behavior, 127
  cerebello-thalamocortical, 121-122
  methodology, 153
  nigro-thalamocortical, 122-125
  pallido-thalamocortical, 122-125
EPSP (excitatory postsynaptic potential), 105-107
  pyramidal tract neurons, 112-113
  in red nucleus, 127-131
Ethology, 40
Eye movement, sporadic, 77

Feedback, set-regulated, 133
Feedback set, 149
Fiber tracing, retrograde *vs.* anterograde, 93
Fiber-tract recording, 153, 154-155

methods, 154–155
weakness of, 154
Field potential(s), 173
  cerebellar-evoked, 118
  of cerebellocortical pathways, 114–121
  in cerebral cortex, 141–151
  components of, 141–142, 145–147
  cross-innervation effects, 129
  defined, 141
  deprivation effects, 145–147
  electrically evoked, 143–145, 148–151
  experimental modification, 145–147
  experiments on set, 148–151
  interpretations, 141–143
  in red nucleus, 127–131
  and short-term behavior changes, 147–148
  synaptic locus, 147
  visual cortical, 143
Flexibility, behavioral, 10, 39–41
Form analysis, of cell types, 167
Frontal agranular cortex:
  cerebellar pathways to, 103–114
  defined, 78
  instruction stimuli effects, 78
  multisynaptic effects, 110
  thalamic inputs to, 91–95
Frontal cortex, 69–78
  neuronal activity, 69
Frontal lobe, 3
Fundus, of intraparietal sulcus, 100
Fusional cells, in striate visual cortex, 167

GABAergic cells, 52
Gag reflex, 32
Gating:
  of cerebellar signals, 137–138
  in corticocortical pathways, 139
  nonselective, 134
  in precentral motor cortex, 136–137
  set-dependent, 137–138
  set-related, 120–121
  sites, 135
  in thalamic nuclei, 137–138
  transmission, 132–133
Globus pallidus:
  electrophysiology, 122–125
  output pathways, 96
  thalamic inputs, 96–99
Golgi neurons, 59

Higher brain function:
  and mediating process, 14
  neurophysiology, 1–3

Horseradish peroxidase, 93

Inferotemporal regions, of cortex, 67
Information processing, 47
  circuits, 49–64
  and columnar organization, 49–50
  intracortical, 52–53
  serial, 52
    vs. parallel, 54
Information theory, 15–16
  and reaction time, 22
Inhibition, monosynaptic, 123
Instruction(s):
  and cell discharge, 72
  visuospatial, 76
Instruction stimulus, 7, 65
  arbitrary, 67
  and motor output, 22
  neuronal activity, 71
  set-related effects, 79
Intelligence, and brain size, 40

Kinesthetic reaction times, 27–29
Kinesthetic stimulus, set-related effects, 79

Laminar analysis, 166–169
  advantage of, 168–169
  of cell stomata, 153
Learning, plasticity vs. flexibility, 3–5
Limbic domain, 42
LP-pulvinar complex, 53

Mammals:
  brain size in, 39–40
  learning ability of, 40
Medial lemniscus, 149
  gating via, 139
Modulation, neuronal patterns, 70–71
Modulatory inputs, to cerebral cortex, 55–56
Monkey cortex:
  cytoarchitectonic maps, 92
  parietal, 58
  pyramidal tract, 45
Motion analysis, 167
Motoneurons, upper vs. lower, 91
Motor cortex, 78–85
  antidromic activation, 158–159
  regions of, 67
  supplementary, 43
Motor set, 21–27
  adaptive plasticity and, 131–134
  and black box, 65–66
  concepts of, 9

Motor set (*Continued*)
  defined, 5
  vs. perceptual set, 8, 17–18
Movement:
  cerebral control of, 44–46
  neuronal model, 12
  reflex *vs.* nonreflex, 25
  set-dependent changes, 21–37
  triggering apparatus, 25
  voluntary, 24
Myotatic reflex, 31

Neocortex:
  and brain size, 39–40
  cell types, 58–59
  function of, 39–40
  in higher brain function, 5, 41
  layers of, 5–6
  organization of, 41–44
Nerves, *see* Neurons
Nervous system, in mammals, 68–69
Neural models, 12
Neural switching, importance of, 4
Neuroanatomy, 40, 53
Neurological disorders, "disconnection syndromes" in, 99
Neurons:
  clusters of, 80
  corticofugal, 168
  corticorubral, 158
  corticospinal, 46
  dentate *vs.* interpositus, 108–109
  directionally specific, 70, 72–73
  enhanced response of, 77
  granular, 59
  lateral geniculate, 52
  "light-sensitive," 86
  local circuit, 60
  mediating process of, 8, 12–15
  modulation patterns, 70
    sustained, 70–71
  monosynaptic excitation of, 105
  output, 1
  pathways of, 91
  plasticity of, 3–4
  projection, 60, 80
  properties of, 42
  pyramidal tract, 79–80, 158
  representation, 174
  set-related, 71, 79
    primitive, 67–69
  size of, 5
  visual cortext, 52
  "visual fixation," 86

Neurophysiology:
  applicability to cortical cell type identification, 169–170
  lower brain function, 2–3
  and pyramidal tract neuron, 1–6
  research strategies, 2–3
Nucleus, interpositus *vs.* dentate, 19. *See also* Red nucleus

Occlusion:
  convergence and, 116–117
  spatial facilitation and, 119
Open-loop afferent system, 34
Optic nerve stimulation, 144, 145

Pallidothalamic projections, 96–97
Parietal cortex, 44
  corticocortical inputs, 99–102
  in monkeys, 100
Parietal lobe, gating via, 139
Parietal somatosensory cortex, 85–86
  neuronal activity, 85
Parietal visual cortex, 86–87
Perception, and expectancy, 9
Perceptual set:
  and black box, 65
  defined, 4
  long-term, 12
  vs. motor set, 17–18, 21
Physiology, early studies, 10–11
Pleurobranchaea, and stimuli suppression, 68
Pons, intrinsic cortical organization, 60
Posterior parietal region, of cortex, 67
Postspike facilitation, 164
Postspike suppression, 164
Precentral motor cortex, 43–44
  functional specialization of, 46–47
  gating in, 136–137
  muscle coordination, 46
  neurons, 44, 46
  set-related activity, 80
Prefrontal cortex, 67, 69–78
  discharge coding, 75
  neuronal activity, 70
Premotor cortex, 44
  neuronal activity, 82–85
Preparatory set:
  and behavior, 7–20
  importance of, 8–9
Preprogramming, 29–30
Primary motor cortex, 149
Primary somatosensory cortex, 149
Psychology, experimental, 9–11

Psychophysics, 15–17
Purkinje protection cells, 59
Pyramidal cells, 51
Pyramidal tract:
 destruction of, 44–45
 electrical stimulation of, 5
 in monkeys, 45
 neocortex layers, 5–6
 neurophysiology, 1–6

Rabbit, visual cortex, 58
Reaction time, 27–29
 choice, 22
 experiment, 23, 24
 kinesthetic vs. auditory, 27–29
 and motor output, 21
 simple, 21
Recording:
 extracellular, 156
 fiber-tract, 153
 intracellular, 155
Red nucleus:
 conditioning stimulus, 128
 EPSP vs. field potential, 127–131
 field potentials of, 128
 interpositus stimulation, 128
 intracellular recordings, 127
 intrinsic cortical organization, 60
 modification by conditioning, 130
 in motor cortex, 58
Reflex(es):
 of marksmen, 31
 neural, 31
 properties of, 10–11
 types of, 32
 and volitional behavior, 30–32
REM (rapid eye movement) sleep, 148
Respiratory movement, coordination, 36

Set:
 and attention, 8
 automatic responses, 33–34
 and "behaviorist revolution," 11
 brain circuits identified for, 134–135
 central program, 149
 defined, 7–8
 dependent behavior, 91–126
 feedback, 133
 field potential experiments, 148–151
 history of, 9–11
 hypothesis formation, 5
 importance of, 7–9
 in information theory, 16
 as mediating process, 10

and motor output, 21–27
neural model, 12, 14
relationship studies, 76–77
research on, 15
and "selective attention," 8
short-term, 14
and signal detection, 16–17
and single unit, 65–89
speech effects, 34–37
switching, 19–20, 127–139
Shadowing, and preparatory set behavior, 18
Sherry-glass response, 33
Short-term set, adaptive plasticity and, 131–134
Signal detection vs. dichotic listening, 18
Single unit:
 neurophysiology, 65
 set and, 65–89
Sleep:
 REM, 148
 slow-wave, 148
 stages, 148
Somata, laminar distribution, 57
Somatic sensorimotor cortex, 42
 in monkey, 57
Spatial instruction, 66–67
Spatial summation, 115
Speech:
 afferent regulation of, 34–37
 experiments, 20
 open-loop systems, 34–35
 and respiratory movement, 36
 set effects on, 34–37
Spike(s):
 collision between, 157
 in field potentials, 141
Spike-triggered averaging, 153, 164–166
 advantages of, 166
 of neuronal cells, 164–166
Spinal cord:
 and behavioral flexibility, 41
 intrinsic cortical organization, 60
 kinesthetic input, 80
States, see Sets
Stellate cells, 50
Stellate neurons, 59
Stimulus-response reactions, 15–16
Stretch reflex, in movement changes, 24
Striate cortex, 52–53
 destruction of, 54
Striatum, intrinsic cortical organization, 60
Subliminal fringe, 116

Substantia nigra, electrophysiology, 122–125
Subtraction set, 15
Supplementary motor cortex, set-related activity, 81–82
Switching:
  between sets, 19–20
  of cerebellar and lemniscal responses, 138
  "dead time" in, 20
  set-related, 127–139
  time-course, 128
  time vs. rate, 19
Synapse, see Axons; Neurons

Temporal summation, 115
Temporal visual cortex, 87
Thalamic cells, retrograde labeling, 94
Thalamic nuclei, in animals, 56
Thalamocortical inputs, to cerebral cortex, 50–52
Thalamocortical projections, 97
  "nonspecific," 125–126
Thalamus:
  cerebellar inputs, 96–99
  cortical inputs, 91–95
  cortical responses, 144
  electrophysiology, 121–125
  globus pallidus inputs, 96–99
  intrinsic cortical organization, 60
  and motor cortex, 19
  neuronal topography, 93–94
  nonspecific nuclei, 125–126
  relay nucleus, 98
  retinal projection, 53
  set-dependent gating in, 137–138
  specific inputs, 125–126
  ventrobasal complex, 139, 149
  ventrolateral, 93, 149
Tonic neuronal discharge, 13
Trigger stimulus, 7, 65
  and motor output, 21
  to muscle shortening, 27

Ventral nuclei, anterior vs. lateral, 92
Vestibulo-ocular reflex, 32, 34
Visual cortex, laminar analysis, 166–168
Visuospatial instruction, 76